Seungkeol Choe

UNIVERSITY OF
UTAH

STOCHASTIC
FINITE ELEMENTS

A *Spectral Approach*

Revised Edition

ROGER G. GHANEM
Johns Hopkins University

POL D. SPANOS
L.B. Ryon Chair in Engineering
Rice University

DOVER PUBLICATIONS, INC.
Mineola, New York

Bibliographical Note

This Dover edition, first published in 2003, is a newly revised edition of the work originally published by Springer-Verlag New York, Inc. in 1991. A new Preface has been prepared especially for this edition.

Library of Congress Cataloging-in-Publication Data

Ghanem, Roger.
    Stochastic finite elements : a spectral approach / Roger G. Ghanem, Pol D. Spanos.—Rev. ed.
        p. cm.
    Includes index.
    ISBN 0-486-42818-4 (pbk.)
    1. Finite element method. 2. Stochastic processes. I. Spanos, P. D. (Pol D.) II. Title.

TA347.F5G56 2003
620'.001'51535—dc21

2003046062

Manufactured in the United States of America
Dover Publications, Inc., 31 East 2nd Street, Mineola, N.Y. 11501

To my Mother and my Father.

R.G.G.

To my parents Demetri and Aicaterine, my first mentors in quantitative thinking; to my wife Olympia, my permanent catalyst in substantive living; and to my children Demetri and Evie, a perpetual source of delightful randomness.

P.D.S.

"... The principal means for ascertaining truth—induction and analogy—are based on probabilities; so that the entire system of human knowledge is connected with the theory (of probability) ... "

Pierre Simon de Laplace,
A Philosophical Essay on Probability, 1816.

"... Nature permits us to calculate only probabilities, yet science has not collapsed."

Richard P. Feynman,
QED: The Strange Theory of Light and Matter, 1985.

Preface to the Dover Edition

Since the publication of the first edition of this book in 1991, the field of stochastic finite elements has greatly benefited from the concepts of spectral representation of stochastic processes in terms of thematic diversity and mathematical foundation. This, in fact, was anticipated in the epilogue of the first printing of the book. In this context, it has been tempting to proceed with a second edition towards incorporating certain of these developments. However, it has been felt that the simplicity and tutorial effectiveness of the original version could be compromised by some of the logistic and conceptual details which would have to be incorporated. Therefore, the authors have decided to proceed with this Dover publication with the hope that the original, and in many respects seminal, concepts would become widely available to broad audience for a longer period of time. It is hoped that in this manner, the many requests that we had for reprinting, worldwide, will be answered.

Thanks are expressed to Dover Publications for accommodating this need and helping disseminate the original contribution towards a rational foundation for uncertainty quantification. Nevertheless, a somewhat exhaustive list of references to the utilization, refinement, and extension of the concepts of this monograph is included as an addendum at the end of the book. It is hoped that this addition will help as a catalyst for even more accelerated use of spectral approaches to new applications and mathematical formulation.

The Authors
September 2002

Preface

This monograph considers engineering systems with random parameters. Its context, format, and timing are correlated with the intention of accelerating the evolution of the challenging field of Stochastic Finite Elements. The random system parameters are modeled as second order stochastic processes defined by their mean and covariance functions. Relying on the spectral properties of the covariance function, the Karhunen-Loeve expansion is used to represent these processes in terms of a countable set of uncorrelated random variables. Thus, the problem is cast in a finite dimensional setting. Then, various spectral approximations for the stochastic response of the system are obtained based on different criteria. Implementing the concept of Generalized Inverse as defined by the Neumann Expansion, leads to an explicit expression for the response process as a multivariate polynomial functional of a set of uncorrelated random variables. Alternatively, the solution process is treated as an element in the Hilbert space of random functions, in which a spectral representation in terms of the Polynomial Chaoses is identified. In this context, the solution process is approximated by its projection onto a finite subspace spanned by these polynomials.

The concepts presented in this monograph can be construed as extensions of the spectral formulation of the deterministic finite element method to the space of random functions. These concepts are further elucidated by applying them to problems from the field of structural mechanics. The corresponding results are found in agreement with those obtained by a Monte-Carlo simulation solution of the problems.

The authors wish to thank Rice University, the Houston Advanced Research Center (HARC), and the National Center for Supercomputing Applications (NCSA) for the extensive use of their computational facilities during the course of the studies which have led to the conception, development, and integration of the material of this monograph. The financial support, over a

viii

period of years, from the National Science Foundation, the National Center for Earthquake Engineering Research at the State University of New York at Buffalo, the Air Force Office of Scientific Research, and Rice University is gratefully acknowledged. Further, the stimulating discussions with a plethora of students and colleagues are greatly appreciated.

R.G. Ghanem
P.D. Spanos
October 1990.

Contents

STOCHASTIC
FINITE ELEMENTS

Chapter 1

INTRODUCTION

1.1 Motivation

Randomness can be defined as a lack of pattern or regularity. This feature can be observed in physical realizations of most objects that are defined in a space-time context. Two sources of randomness are generally recognized (Matheron, 1989). The first one is an inherent irregularity in the phenomenon being observed, and the impossibility of an exhaustive deterministic description. Such is the case, for instance, with the Uncertainty Principle of quantum mechanics and the kinetic theory of gas. The other source of randomness can be related to a generalized lack of knowledge about the processes involved. The level of uncertainty associated with this class of problems can usually be reduced by recording more observations of the process at hand and by improving the measuring devices through which the process is being observed. In this category fall, for instance, the econometric series whereby the behavior of the financial market is modeled as a stochastic process. This process, however, can be entirely specified, in the deterministic sense, from a complete knowledge of the flow of goods at the smallest levels. Another example that is of more direct concern in the present study pertains to the properties of a soil medium. These properties are uniquely defined at a given spatial location within the medium. It is quite impractical, however, to measure them at all points, or even at a relatively large number of points. From a finite number of observations, these properties may be modeled as random variables or, with a higher level of sophistication, as random processes with the actual medium properties viewed as a particular realization of these processes. From the preceding it is apparent that deterministic models can be considered as approximations to the corresponding physical problems, in a similar sense that linear models are approximations to the actual nonlinear behavior of these problems.

The randomness in the response of a system can be induced either by the input or the system operator. From a causality perspective, the output cannot exist without the presence of an excitation or some initial conditions on the operator. A parallel probabilistic argument can be made to the extent that the output cannot be random without either the input or the operator being random.

The case of a random input and a deterministic operator has been amply studied and numerous results have been obtained in the field of engineering mechanics. The second case, that of a random operator, with or without random inputs, is much more complicated and the underlying mathematical tools are still in the developing stage. This operator induced randomness is also termed parametric. Within this class of problems, the degree of complexity depends again on the extent of sophistication required from the model. A major factor in this respect is whether the random aspect of the problem is modeled as a random variable or as a random process. Obviously, the requisite accuracy and hence sophistication of the model depends on the relative importance of the system being investigated. This importance factor is related to such concepts as the generalized cost to be incurred from a potential failure.

With the startling technological growth witnessed during recent decades, the development of accurate and realistic models of engineering systems has become feasible. The functionality of many modern structural systems depends to a large extent on their ability to perform adequately and with a high level of reliability under not absolutely controllable conditions. Such structures include, among others, nuclear power plants, offshore platforms, high-rise buildings, and reentry space vehicles. The randomness in these cases results from such diverse phenomena as soil variability, earthquake-induced ground motion, random ocean waves, thermal and acoustic loadings, and uncertain fatigue effects. These are examples of important structures that should be designed with a negligible probability of failure. In addition to being invaluable for the safe design of such structures, the probabilistic approach provides crucial concepts for the task of code development for general engineering practice.

Paralleling this technology induced option for more sophisticated models, basic theoretical developments have been underway in relation to the analysis of random processes. These developments were triggered by the need for more accurate models in theoretical statistics and electrical engineering. More will be said in Chapter II concerning that portion of the theory that is directly related to the theme of this monograph, namely the representation theory for random processes.

Random aspects aside, modern engineering structures are quite complex to analyze. Such complexity results from the intricate interaction among the different parts of the structure, as well as the interaction of the structure with the surrounding medium. Addressing such complexity requires recourse to accurate and efficient numerical algorithms. Therefore, it is clear that for any probabilistic modeling of a physical system to be useful, it has to be compatible with the available numerical schemes. The finite element method has proven to be well suited for a large class of engineering problems. Finite element algorithms have a sound and well developed theoretical basis and their efficiency has long been proven and tested on a variety of problems. The formulation presented herein is, indeed, a natural extension of the basic ideas of the deterministic finite element method to accommodate random functions.

1.2 Review of Available Techniques.

From an engineering mechanics perspective, the most common stochastic system problem involves a linear differential equation with random coefficients. These coefficients represent the properties of the system under investigation. They can be thought of as random variables or, more accurately and with an increasing level of complexity, as random processes with a specified probability structure. Mathematically, the problem can be formulated as

$$\Lambda\, u \,=\, f \qquad\qquad (1.1)$$

where Λ is a linear stochastic differential operator, u is the random response, and f is the possibly random excitation. Note that equation (1.1) with a deterministic operator Λ and a random excitation f has been studied extensively (Y.K. Lin, 1967; Yang and and Kapania, 1984; Y.K. Lin et.al, 1986). The case where the operator Λ is stochastic is considerably more difficult and only approximate solutions to the problem have been reported in the literature. A rigorous mathematical theory has been developed for the solution process when equation (1.1) is of the Itô type (Itô, 1951). In this case, the solution is a Markov process the probability distribution of which satisfies a Fokker-Planck, partial differential, equation which in general is not amenable to analytical solutions. Implicit in the Itô equation is the assumption that the random coefficients associated with the operator are white noise processes. The relationship between ordinary equations and equations of the Itô type was considered by Wong and Zakai (1965). The stability of linear stochastic equations with Poisson process coefficients was also investigated (Li and Blankenship, 1986). Along similar lines, and through a judicious use of the relationship between diffusion and random-walk, the Navier-Stokes equation for flows with high Reynolds numbers was numerically analyzed by Chorin (1973). The approach which was used featured a grid-free numerical solution of the governing equations, which would otherwise require a prohibitively fine discretization. This solution strategy hinges on an analogy between the diffusive part of the Navier-Stokes equation and random-walk, thus permitting the simulation of the process of vorticity generation and dispersal by using computer-generated pseudo-random numbers. Various convergence studies of this method have since been carried out (Roberts, 1989), along with a number of applications, specially in connection with turbulence related problems (Ghoniem and Oppenheim, 1984). A more realistic problem is that of a stochastic operator the random coefficients of which possess smooth sample paths. For this class of problems the Itô calculus breaks down, and no general mathematical theory is available for obtaining exact solutions. Kozin (1969) presented a comprehensive review of the stability problem of these equations as well as equations of the Itô type. Bharrucha-Reid (1959) and Hans (1961) were among the first to investigate the problem from a mathematical point of view. They studied the existence and uniqueness of the solution to random operator equations.

The problem in dealing with stochastic equations is two-fold. Firstly, the random properties of the system must be modeled adequately as random variables or processes, with a realistic probability distribution. A good treatment of this modeling phase is presented by Benjamin and Cornell (1970), and by Vanmarcke (1983). Also, a compact probabilistic representation has been suggested by Masri and Miller (1982), and by Traina et.al (1986). Secondly, the resulting differential equation must be solved, and response quantities of interest are obtained, usually as determined by their second order statistics. It is this second stage which is of immediate concern in the present study. Namely, the solution of equation (1.1) when the coefficients of the operator Λ possess a prescribed probabilistic structure. Interest in this class of differential equations has its origins in quantum mechanics, wave propagation, turbulence theory, random eigenvalues, and functional integration. Good accounts of these applications are presented by Bharrucha-Rei(1968) and by Sobczyk (1985). The methods of solution range from solving the averaged equations for an equivalent solution, to perturbing the random fields and keeping one or two terms in the expansion, or using a Born approximation for the random fields. Some attempts have also been directed at obtaining exact solutions to the problem by functional integration (Hopf, 1952). He suggested the construction of a differential equation for the characteristic functional of the random solution, from which a complete probability description can be obtained. The method was applied by Hopf to a turbulence problem. Lee (1974) applied functional integration to the problem of wave propagation in random media, and obtained a Fokker-Planck equation for the characteristic functional of the solution. Later, Kotulski and Sobczyk (1984) developed a scheme to construct the characteristic functional of the stochastic evolution equation and obtained expressions that could be evaluated only under very restrictive conditions. The characteristic functional approaches tend, in general, to be of a qualitative and specialized nature.

From the above discussion it becomes clear that the current state of knowledge concerning stochastic differential equations precludes exact solutions except in some rare instances. The problem becomes even more intractable when applying the results described above to engineering problems with intricate geometries and boundary conditions and various types of excitations. Comprehensive reviews of the current trends for analyzing random systems in engineering are given by Iyengar and Dash (1976), Ibrahim (1987), Shinozuka (1987b) and Benaroya and Rehak (1988).

The most widely used technique for analyzing random systems in engineering is the perturbation method. This fact is mainly due to the mathematical simplicity of the method, which does not, however, guarantee its validity as a method of solution. The perturbation scheme consists of expanding all the random quantities around their respective mean values via a Taylor series. Secular terms that cause instability of the approximate

solution appear in higher order terms. Also, computations beyond the first
or second terms are not practical. These facts restrict the applicability
of the method to problems involving small randomness, as witnessed in
the available literature. Boyce and Goodwin (1964) used the perturbation
approach for the solution of the eigenvalue problem of random strings and
beams. They used the integral equation approach to invert the governing
differential operator, and considered randomness caused both by material
properties and boundary conditions. Soong and Bogdanoff (1963) used
transfer matrix techniques to investigate the behavior of disordered linear
chains. They expressed the frequency response of the chain as a perturbation
type expansion in terms of the small deviations of the random variables
involved. They concluded that even for small levels of disorder in the system,
the higher frequencies showed considerable spread around their mean values.
Later, Collins and Thompson (1969) used the perturbation approach to
treat the general problem of computing eigenvalue and eigenvector statistics
of a system whose parameters were random variables described by their
covariance matrix. A second order Taylor series about the mean value was
used to represent the eigenvalues, eigenvectors and the random parameters
of the system. They also investigated the stability problem of a column
and concluded that stability problems were more sensitive to cross-sectional
area uncertainty than frequency problems. Hart and Collins (1970) and
Hasselman and Hart (1972) extended the work to develop a perturbation
scheme compatible with the finite element method. Dendrou and Houstis
coupled results from an inference statistical analysis of a soil medium (1978a)
with the finite element method to solve a soil-structure interaction problem
(1978b). They used a first order perturbation scheme which again, restricted
their approach to rather small levels of random fluctuations. Using the same
perturbation type expansion, Nakagiri and Hisada (1982) and Hisada and
Nakagiri (1985) investigated again the random eigenvalue problem, allowing
for random boundary conditions. They concluded that the second order
perturbation was too untractable to be of any practical interest in solving
real physical problems. Liu et.al (1985) introduced a new implementation
scheme for the perturbation based finite element method. The perturbation
expansion was carried out at a different stage of the finite element assembly
process. The resulting equations, however, are identical with those obtained
from the more standard method mentioned previously. Liu et.al applied their
approach to analyzing random nonlinear and transient problems (1985, 1986,

1988).

All the methods discussed above are based on a perturbation type expansion of the random quantities involved. Their validity is restricted to cases where the random elements exhibit small fluctuations about their mean values, in which case a first order approximation yields fairly good results. Also, a solution to the problem based on the perturbation approach does not provide, without a prohibitively large amount of additional analytical and computational effort, higher order statistics.

Another more elegant method, not very popular in connection with the class of problems treated herein, is the hierarchy closure approximation (Bharrucha-Reid, 1968). The method is based on expressing joint statistical moments of the output and of the system as functions of lower order moments. Consider, for example, equation (1.1) where the stochastic operator Λ is split into a deterministic part $\mathbf{L}$ and a random part $\mathbf{\Pi}$

$$(\mathbf{L} + \mathbf{\Pi})\, u \, = \, f \, . \tag{1.2}$$

Solving for u results in

$$u \, = \, \mathbf{L}^{-1} f \, - \, \mathbf{L}^{-1} \mathbf{\Pi} u \, . \tag{1.3}$$

Note that on averaging equation (1.3), the moments of u depend on those of $\mathbf{\Pi} u$. To obtain these latter moments, apply the operator $\mathbf{\Pi}$ to equation (1.3), to derive

$$\mathbf{\Pi} u \, = \, \mathbf{\Pi} \mathbf{L}^{-1} f \, - \, \mathbf{\Pi} \mathbf{L}^{-1} \mathbf{\Pi} u \, . \tag{1.4}$$

Substituting equation (1.4) into equation (1.3), gives

$$u \, = \, \mathbf{L}^{-1} f \, - \, \mathbf{L}^{-1} \mathbf{\Pi} \mathbf{L}^{-1} f \, + \, \mathbf{L}^{-1} \mathbf{\Pi} \mathbf{L}^{-1} \mathbf{\Pi} u \, , \tag{1.5}$$

and so on. At some point in the substitution process, an approximation has to be made of the form

$$< \left(\mathbf{L}^{-1} \mathbf{\Pi} \right)^{n} u > \, = \, < \left(\mathbf{L}^{-1} \mathbf{\Pi} \right)^{n} > <u> \, , \tag{1.6}$$

where $<.>$ denotes the operation of mathematical expectation. The uncoupling in equation (1.6) has no rigorous basis. It is often justified intuitively by a local independence argument. It was suggested by Adomian (1983) that closure at a certain level of the hierarchy is equivalent to the same order of perturbation in a perturbation based solution. The difficulties in formulating higher order closure approximations are substantial enough to limit the applicability of the method to cases of small random fluctuations.

Vanmarcke and Grigoriu (1983) presented a finite element analysis of a simple shear beam with random rigidity. Closed form solutions were obtained for the statistics of the response. However, the very restrictive assumptions concerning the characteristics of the beam make the method of a rather limited interest.

Adomian (1964) introduced the concept of stochastic Green's function for the solution of differential equations involving stochastic coefficients. The concept was further developed by Adomian and Malakian (1980) and Adomian (1983) who suggested using a decomposition method for solving nonlinear stochastic differential equations. Accordingly, the inverse of the random operator is expanded in a Neumann series (Courant and Hilbert, 1953), and the random resolvent kernel of the problem is computed. The result can be viewed as a series approximation to the random solution process. Obtaining higher order moments for the solution process is quite cumbersome since it involves averaging products of random matrices. The method was applied by Benaroya (1984) and Benaroya and Rehak (1987) for the solution of single-degree-of-freedom structural dynamics problems. Good results were obtained for small values of the coefficient of variation of the random elements. However, and even with the assistance of MACSYMA (1986), a symbolic manipulation package, difficulties were experienced in including more than two terms in the expansion, a fact that handicaps the method when dealing with sizable randomness. Further, Shinozuka and Nomoto (1980) developed the delta method, again a Neumann expansion based solution, for solving problems with random media. They implemented a first order approximation which gave acceptable results only for the case of very small fluctuations. Later, Yamazaki et.al (1985) introduced the Monte Carlo Expansion (MCE) method, which constitutes a blend of the Neumann expansion with the Monte Carlo simulation. The method was subsequently applied by Shinozuka and Deodatis (1986) for the solution of problems involving random plates and beams. It is based on the Neumann expansion for the inverse of the stochastic coefficient matrix whereby digital simulation techniques are used to generate the random matrix. The method was applied to problems with large coefficients of variation and good agreement was observed with the standard Monte Carlo simulation. However, the method requires simulating the random field several times in order to produce reliable results, especially when sizable random fluctuations are involved.

1.3 The Mathematical Model

The class of problems dealt with in this study is not of the conventional engineering kind in that it involves concepts of a rather abstract and mathematical nature. It is both necessary and instructive to introduce at this point the mathematical concepts which are used in the sequel. The degree of abstraction is kept to a necessary minimum to avoid obscuring the engineering aspect of the problem. Also a notation is introduced that simplifies, to some extent, the mathematical complexity. It may seem, at first, that a mathematical abstraction of the problem is gratuitous. It is believed, however, that such an approach is vital for a complete and mature understanding of the problem as well as directing any future research aimed at extending the present formulation.

The Hilbert space of functions (Oden, 1979) defined over a domain $\mathbf{D}$, with values on the real line R, is denoted by $\mathbf{H}$. Let $(\mathbf{\Omega}, \mathbf{\Psi}, P)$ denote a probability space. Let $\mathbf{x}$ be an element of $\mathbf{D}$ and θ be an element of $\mathbf{\Omega}$. Then, the space of functions mapping $\mathbf{\Omega}$ onto the real line is denoted by $\mathbf{\Theta}$. Each map $\mathbf{\Omega} \to R$ defines a random variable. Elements of $\mathbf{H}$ and $\mathbf{\Theta}$ are denoted by roman and greek letters respectively. Capital letters are used to denote algebraic structures and spaces as well as operators defined on these spaces, with greek letters referring again to those operators defined on spaces of random functions.

The inner products over $\mathbf{H}$ and over $\mathbf{\Theta}$ are defined using the Lebesgue measure and the probability measure, respectively. That is, for any two elements $h_i(\mathbf{x})$ and $h_j(\mathbf{x})$ in $\mathbf{H}$, their inner product $(\ h_i(\mathbf{x})\ ,\ h_j(\mathbf{x})\ )$ is defined as

$$(h_i(\mathbf{x}), h_j(\mathbf{x})) = \int_{\mathbf{D}} h_i(\mathbf{x}) h_j(\mathbf{x}) d\mathbf{x} . \tag{1.7}$$

The domain $\mathbf{D}$ represents the physical space over which the problem is defined. Similarly, given any two elements $\alpha(\theta)$ and $\beta(\theta)$ in $\mathbf{\Theta}$, their inner product is defined as

$$(\alpha(\theta), \beta(\theta)) = \int_{\mathbf{\Omega}} \alpha(\theta) \beta(\theta) dP \tag{1.8}$$

where dP is a probability measure. Under very general conditions, the integral in equation (1.8) is equivalent to the average of the integrand with respect to the probability measure dP, so that

$$(\alpha(\theta), \beta(\theta)) \;=\; \alpha(\theta)\beta(\theta) \;. \tag{1.9}$$

Any two elements of the Hilbert spaces defined above are said to be orthogonal if their inner product vanishes. A random process may then be described as a function defined on the product space $\mathbf{D} \times \Omega$. Viewed from this perspective a random process can be regarded as a curve in either of $\mathbf{H}$ or $\boldsymbol{\Theta}$.

The physical model under consideration involves a medium whose properties exhibit random spatial fluctuations and which is subjected to a random external excitation. The mathematical representation of this problem involves an operator equation

$$\boldsymbol{\Lambda}(\mathbf{x}, \theta)u(\mathbf{x}, \theta) \;=\; f(\mathbf{x}, \theta) \tag{1.10}$$

where $\boldsymbol{\Lambda}(\mathbf{x}, \theta)$ is some operator defined on $\mathbf{H} \times \boldsymbol{\Theta}$. In other words, $\boldsymbol{\Lambda}$ is a differential operator with coefficients exhibiting random fluctuations with respect to one or more of the independent variables. The aim then is to solve for the response $u(\mathbf{x}, \theta)$ as a function of both its arguments. With no loss of generality, $\boldsymbol{\Lambda}$ is assumed to be a differential operator, whose random coefficients are restricted to being second order random processes. This is not a severe restriction for practical problems, since most physically measurable processes are of the second order type. Then, each one of these coefficients $a_k(\mathbf{x}, \theta)$ can be decomposed into a purely deterministic component and a purely random component in the form

$$a_k(\mathbf{x}, \theta\,) \;=\; \bar{a}_k(\mathbf{x}) \;+\; \alpha_k(\mathbf{x}, \theta) \tag{1.11}$$

where $\bar{a}_k(\mathbf{x})$ is equal to the mathematical expectation of the process $a_k(\mathbf{x}, \theta)$, and $\alpha_k(\mathbf{x}, \theta)$ is a zero-mean random process, having the same covariance function as the process $a_k(\mathbf{x}, \theta)$. Equation (1.10) can then be written as

$$[\mathbf{L}(\mathbf{x}) \;+\; \boldsymbol{\Pi}(\mathbf{x}, \theta)]\,u(\mathbf{x}, \theta) \;=\; f(\mathbf{x}, \theta) \;, \tag{1.12}$$

where $\mathbf{L}(\mathbf{x})$ is a deterministic differential operator and $\boldsymbol{\Pi}(\mathbf{x}, \theta)$ is a differential operator whose coefficients are zero-mean random processes. Before a solution to equation (1.12) is sought, it is essential to clarify what is meant by such a solution.

As was mentioned above, Θ denotes the Hilbert space of functions defined on the $\sigma-$field of events generated by the physical problem, with range the interval $[0, 1]$. In other words, if the possible realizations of the random process were numbered continuously on the interval $[0, 1]$, and these numbers were assigned to the variable θ, then, for a fixed $\theta = \theta^* \in [0, 1]$, the process $\alpha_k(\mathbf{x}, \theta^*)$ is a deterministic function of $\mathbf{x}$, a realization of the process. From observing a finite number of realizations of the process, distribution theory can be used to construct the distribution of the process along the θ dimension. For a given $\mathbf{x}$, $a_k(\mathbf{x}, \theta)$ is a random variable with such a distribution. Obviously, for a complete description of the process, the joint distribution at all $\mathbf{x} \in \mathbf{D}$ is required. However, if the process is assumed to be Gaussian, all the finite dimensional distribution functions are also Gaussian. Clearly, the usually limited number of observed realizations of a random process cannot, in general, suggest any definite distribution. However, invoking the central limit theorem, the Gaussian distribution appears to be the most likely candidate for many physical applications.

Once the coefficients in the operator equation (1.10) have been defined through their probability distribution functions, the main question remains as to what is meant by a solution to the problem. Obviously, the discussion and comments just made concerning the coefficients processes apply to the solution process as well. A conceptual modification, though, can be introduced. Specifically, a quite general form of the solution process can be expressed as

$$u = g\left[\alpha_k(\mathbf{x}, \theta), f(\mathbf{x}, \theta), \mathbf{x}\right] \tag{1.13}$$

where $g[.]$ is some nonlinear functional of its arguments. Clearly, a complete description of the response would involve the prescription of its joint distribution with the various processes appearing in equation (1.13). This information could form the basis for a rational reliability and risk assessment. However, given the infinite dimensional structure of the random processes appearing in equation (1.13), such a task seems to exceed the capability of currently used methods. A finite-dimensional description of the processes involved is required if the solution is to proceed in a computational setting. Given the abstract nature of the functional spaces over which random processes are defined, a finite dimensional representation cannot be achieved through partitioning of these spaces as is usually done with the deterministic finite element analysis. Alternatively, an abstraction of the discretization process can be introduced which is mathematically equivalent to a discretization with respect to a spectral measure. Indeed, a number of spectral representations are introduced in the text which permit the algebraic manipulation of random processes through that of an equivalent discrete set of random variables. As the title indicates, this monograph is a study in the application of spectral discretizations, taken in the context just explained, to problems of mechanics treatable by the finite element method. This is, indeed, the main impetus of the book.

1.4   Outline

In Chapter I, available techniques for addressing problems with stochasticity are reviewed. Further, mathematical concepts useful in the ensuing analysis are presented.

In Chapter II, relevant elements of the representation theory for continuous stochastic processes are presented. The theory is reviewed and the two representations used in the ensuing development of the stochastic finite element method are discussed in detail. These are the Karhunen-Loeve expansion and the Homogeneous Chaos expansion.

Chapter III includes a review of deterministic finite element methods emphasizing the features that can be extended to the stochastic case. Further, a number of the most widely used techniques for treating the model developed in this chapter are reviewed. Attention is focused on methods based on the spectral representation of stochastic processes as outlined in Chapter II. These methods are shown to provide a sound theoretical extension of the deterministic finite element method to problems involving random system properties.

In Chapter IV reliability methods in structural engineering are briefly reviewed. Further, it is shown how the developments of the previous chapters can be used in a number of approaches to compute approximations for the probability density function of the response of the system.

In Chapter V detailed numerical examples are given which demonstrate the usefulness of the spectral approaches introduced in Chapter III.

Finally in Chapter VI, pertinent conclusions and a broad perspective of the problem at hand are presented.

Chapter 2

REPRESENTATION OF STOCHASTIC PROCESSES

2.1 Preliminary Remarks

Similarly to the case of the deterministic finite element method, whereby functions are represented by a denumerable set of parameters consisting of the values of the function and its derivatives at the nodal points, the problem encountered in the stochastic case is that of representing a random process by a denumerable set of random variables, thereby discretizing the process. In a more applied sense, it is sought to describe random processes in such a manner that they can be implemented in a finite element formulation of the physical problem.

In the deterministic case discretization of the domain has a physical appeal. The domain in the stochastic case does not, however, have a physical meaning that permits a sensible discretization. In this context the abstract Hilbert space foundation of the finite element method becomes useful as it can be extended to deal with random functions. The requisite mathematical rigor for a complete understanding of the theory of representation for stochastic processes is beyond the scope of this book. However, a brief review of the theory will provide more insight into the problem at hand and a broad perspective of the relevant mathematical concepts. In this context, the heuristic local averaging representation, and the more rigorous spectral representation are reviewed in the next section.

2.2   Review of the Theory

Although rather heuristic, local averaging is probably the most widely used method for representing random processes. A continuous random process is formally defined as an indexed set of random variables, the index belonging to some continuous uncountable set. Then, the process can be approximated as closely as desired by restricting the index to a set dense in the indexing set. Stretching mathematical rigor further, a random process can be represented by its values at a discrete set of points in its domain of definition. It is then clear how the foregoing discussion for random variables can be extended to the case of random processes. In the context of the finite element analysis of a given system, the random processes involved are substituted for by random variables that are so chosen as to coincide with some local average of the process over each element. It is expected, however, that the result would depend to a notable extent on the averaging method used. Local averages are usually of two kinds. These are the weighted average over a subset of the indexing set, and the collocation average over each such subset whereby the process over the subset is replaced by its value at some point in the subset. It is clear that, in general, the first approach smoothes the random process, whereas the second approach introduces additional irregularities. This suggests that the two approaches provide lower and upper bounds, respectively, for the random behavior of the process. In the limit, as the size of each subset becomes vanishingly small, the representation resulting from the two approaches should converge to the exact process. It is obvious that a relatively large number of random variables is required to represent a random process in this fashion. It is noted that local averaging parallels pointwise approximations of deterministic functions. The size of the individual subsets used for the averaging process does in general depend on the frequency content of the process. That is, the broader the frequency content of the process, the smaller the region over which the process shows a definite pattern, and thus the smaller the size of the necessary subset to meet a certain precision criterion. This problem with local averaging is particularly crucial in the context of the finite element analysis of structures with curved and irregular geometry. The finite element analysis of these systems usually requires recourse to curved elements and non-uniform spacing of the nodal points. The shape and size of the finite element mesh is generally dictated by the stress distribution within the structure. This stress distribution is usually independent of that of the uncertainty of the random parameters involved. This fact necessitates either the use of an independent mesh for the simulation, or the use of a mesh size such that both stresses and material properties are adequately and consistently represented. In either case, the dimension of the problem, as reflected by the number of random variables used to represent the underlying processes, is quite large. The associated computational problem is, in general, of prohibitive dimensions.

Alternatively to the heuristic arguments associated with the local averaging approach, a rigorous exposition of the basic concepts of the theory of representation for random processes can be formulated (Parzen, 1959). This theory is a quite rich and mature mathematical subject. The development of the theory parallels that of the modern theory of random processes, and has had its origin in the need for more sophisticated models in applied statistics. Most of the related results have been derived for the class of second order processes. Perhaps the most important result is the spectral representation of random processes (Gel'fand and Vilenkin, 1964), which, in its most general form, can be stated as

$$w(\mathbf{x}, \theta) = \int g(\mathbf{x}) \, d\mu(\theta) \tag{2.1}$$

where $w(\mathbf{x}, \theta)$ is a stochastic process whose covariance function $C_{ww}(\mathbf{x}_1, \mathbf{x}_2)$ admits of the decomposition

$$C_{ww}(\mathbf{x}_1, \mathbf{x}_2) = \int g(\mathbf{x}_1) \, g(\mathbf{x}_2) \, <d\mu_1(\theta) \, d\mu_2(\theta)> \, . \tag{2.2}$$

In equation (2.1), $g(\mathbf{x})$ is a deterministic function. Further, $d\mu(\theta)$ is an orthogonal set function, also termed orthogonal stochastic measure, defined on the $\sigma-$field Ψ of random events. An important specialization of the spectral decomposition occurs if the process $w(\mathbf{x}, \theta)$ is wide sense stationary. In this case, equation (2.1) can be shown to reduce to the Wiener-Khintchine relation (Yaglom, 1962) and the following equations hold

$$w(\mathbf{x}, \theta) = \int_{-\infty}^{+\infty} e^{i\mathbf{x}^T.\boldsymbol{\omega}} \, d\mu(\boldsymbol{\omega}, \theta) \tag{2.3}$$

and

$$C_{ww}(\mathbf{x}_1, \mathbf{x}_2) = \int_{-\infty}^{+\infty} e^{i(\mathbf{x}_1 - \mathbf{x}_2)^T.\boldsymbol{\omega}} \, S(\boldsymbol{\omega}) \, d\boldsymbol{\omega}^T \, . \tag{2.4}$$

Here, the symbol T denotes vector transposition, $S(\boldsymbol{\omega})$ is the usual spectral density of the stationary process, and $\boldsymbol{\omega}$ is the wave number vector.

The preceding representations have had a strong impact on the subsequent development of the theory of random processes. However, their applications have been restricted to randomly excited deterministic systems. This is largely attributed to the fact that all of these representations involve differentials of random functions, and are therefore set in an infinite dimensional space, not readily amenable to computational algorithms. An alternative formulation of the spectral representation, and one which is extensively used in the sequel, is the Karhunen-Loeve expansion whereby a random process $w(\mathbf{x}, \theta)$ can be expanded in terms of a denumerable set of orthogonal random variables in the form

$$w(\mathbf{x}, \theta) \;=\; \sum_{i=1}^{\infty} \mu_i(\theta)\, g_i(\mathbf{x}) \;. \tag{2.5}$$

where $\{\mu_i(\theta)\}$ is a set of orthogonal random variables and $g_i(\mathbf{x})$ are deterministic functions, which can be related to the covariance kernel of $w(\mathbf{x}, \theta)$. Note that since equation (2.5) constitutes a representation of the random process in terms of a denumerable set of random variables, it may be regarded as an abstract discretization of the random process. Further, it is important to note that this equation can be viewed as a representation of the process $w(\mathbf{x}, \theta)$ as a curve in the Hilbert space spanned by the set $\{\xi_i(\theta)\}$. The random process $w(\mathbf{x}, \theta)$ is expressed as a direct sum of orthogonal projections in this Hilbert space whereby the magnitudes of the projections on successive basis vectors are proportional to the corresponding eigenvalues of the covariance function associated with the process $w(\mathbf{x}, \theta)$. Given the crucial role that the Karhunen-Loeve expansion has in relation to the methods discussed in this monograph, it will be treated in greater detail in section (2.3).

Collectively, the representations discussed thus far can be thought of as linear operators or filters acting on processes with independent increments (Doob, 1953). Interestingly, these concepts can be generalized to allow for the representation of nonlinear functionals of the orthogonal stochastic measures $d\mu(\theta)$. The theory of nonlinear functionals was developed by Volterra (1913). He generalized the Taylor expansion of functions to the case of functionals. It was Wiener, however, who first applied Volterra's ideas to stochastic analysis, and developed what is now known as the Homogeneous Chaos. Based on Wiener's work, Cameron and Martin (1947) developed the Fourier-Hermite expansion, which is a Fourier-type expansion for nonlinear functionals. Again, it was Wiener (1958) who first applied the new theory to problems involving random phenomena, using the ideas he had developed on Differential Spaces (1923) and Homogeneous Chaos (1938). The resulting expansion, the Wiener-Hermite expansion, was applied by a number of researchers to a large variety of problems. In his development of the theory, Wiener used the Wiener process, which is by definition Gaussian, as a basis for expanding the functionals. Subsequent researchers attempted to generalize Wiener's work to accommodate functionals of other processes (Ogura 1972; Segall and Kailath, 1976). Note that in contrast to the well established linear theory, where a large number of tools are available to the analyst, the nonlinear theory is quite recent and much more intricate to deal with. However, it does hold the promise for substantial extensions of both theoretical and applied nature. An application oriented treatment of the Volterra-Wiener theory for nonlinear systems can be found in the monographs by Schetzen (1980) and Rugh (1981). More recent applications of the Volterra and Wiener-Hermite expansions can be found in the field of time series analysis. Specifically, bilinear time series models (Rao and Gabr, 1984) have been shown to be closely related to the Volterra series expansion (Brockett, 1976). Also, state-dependent-models (SDM), (Priestley, 1988) are based on replacing the infinite-dimensional process by one with finite memory, and identifying the Volterra or Wiener-Hermite expansion of the resulting process.

2.3 Karhunen-Loeve Expansion

2.3.1 Derivation

One of the major difficulties associated with the numerical incorporation of random processes in finite element analyses, is the necessity to deal with abstract measure spaces that have limited physical intuitive support. The major conceptual difficulty from the viewpoint of the class of problems considered herein, involves the treatment of functions defined on these abstract spaces, namely random variables defined on the σ-field of random events. The most widely used method, the Monte Carlo simulation, consists of sampling these functions at randomly chosen elements of this σ-field, in a random, collocation-like, scheme. Obviously, a quite large number of points needs to be sampled if a good approximation is to be achieved. A theoretically more sound and more appealing approach would be to expand these functions in a Fourier-type series as

$$w(\mathbf{x}, \theta) = \sum_{n=0}^{\infty} \sqrt{\lambda_n} \xi_n(\theta) \, f_n(\mathbf{x}) , \qquad (2.6)$$

where $\{\xi_n(\theta)\}$ is a set of random variables to be determined, λ_n is some constant, and $\{f_n(\mathbf{x})\}$ is an orthonormal set of deterministic functions. This is exactly what the Karhunen-Loeve expansion achieves. The expansion was derived independently by a number of investigators (Karhunen, 1947; Loeve, 1948; Kac and Siegert, 1947).

Let $w(\mathbf{x}, \theta)$ be a random process, function of the position vector $\mathbf{x}$ defined over the domain $\mathbf{D}$, with θ belonging to the space of random events $\boldsymbol{\Omega}$. Let $\bar{w}(\mathbf{x})$ denote the expected value of $w(\mathbf{x}, \theta)$ over all possible realizations of the process, and $C(\mathbf{x}_1, \mathbf{x}_2)$ denote its covariance function. By definition of the covariance function, it is bounded, symmetric and positive definite. Thus, it has the spectral decomposition (Courant and Hilbert, 1953)

$$C(\mathbf{x}_1, \mathbf{x}_2) = \sum_{n=0}^{\infty} \lambda_n \, f_n(\mathbf{x}_1) \, f_n(\mathbf{x}_2) \qquad (2.7)$$

where λ_n and $f_n(\mathbf{x})$ are the eigenvalue and the eigenvector of the covariance kernel, respectively. That is, they are the solution to the integral equation

$$\int_{\mathbf{D}} C(\mathbf{x}_1, \mathbf{x}_2) \, f_n(\mathbf{x}) \, d\mathbf{x}_1 = \lambda_n \, f_n(\mathbf{x}_2) . \qquad (2.8)$$

Due to the symmetry and the positive definiteness of the covariance kernel (Loeve, 1977), its eigenfunctions are orthogonal and form a complete set. They can be normalized according to the following criterion

$$\int_{\mathbf{D}} f_n(\mathbf{x}) \, f_m(\mathbf{x}) \, d\mathbf{x} = \delta_{nm} , \qquad (2.9)$$

where δ_{nm} is the Kronecker delta. Clearly, $w(\mathbf{x}, \theta)$ can be written as

$$w(\mathbf{x}, \theta) = \bar{w}(\mathbf{x}) + \alpha(\mathbf{x}, \theta) , \qquad (2.10)$$

where $\alpha(\mathbf{x}, \theta)$ is a process with zero mean and covariance function $C(\mathbf{x}_1, \mathbf{x}_2)$. The process $\alpha(\mathbf{x}, \theta)$ can be expanded in terms of the eigenfunctions $f_n(\mathbf{x})$ as

$$\alpha(\mathbf{x}, \theta) = \sum_{n=0}^{\infty} \xi_n(\theta) \sqrt{\lambda_n} \, f_n(\mathbf{x}) . \qquad (2.11)$$

Second order properties of the random variables ξ_n can be determined by multiplying both sides of equation (2.11) by $\alpha(\mathbf{x}_2, \theta)$ and taking the expectation on both sides. Specifically, it is found that

$$
\begin{aligned}
C(\mathbf{x}_1, \mathbf{x}_2) &= <\alpha(\mathbf{x}_1, \theta)\, \alpha(\mathbf{x}_2, \theta)> \qquad\qquad (2.12) \\
&= \sum_{n=0}^{\infty} \sum_{m=0}^{\infty} <\xi_n(\theta)\, \xi_m(\theta)> \sqrt{\lambda_n\, \lambda_m}\, f_n(\mathbf{x}_1)\, f_m(\mathbf{x}_2) \; .
\end{aligned}
$$

Then, multiplying both sides of equation (2.12) by $f_k(\mathbf{x}_2)$, integrating over the domain $\mathbf{D}$, and making use of the orthogonality of the eigenfunctions, yields

$$
\begin{aligned}
\int_{\mathbf{D}} C(\mathbf{x}_1, \mathbf{x}_2)\, f_k(\mathbf{x}_2)\, d\mathbf{x}_2 &= \lambda_k\, f_k(\mathbf{x}_1) \qquad\qquad (2.13) \\
&= \sum_{n=0}^{\infty} <\xi_n(\theta)\, \xi_k(\theta)> \sqrt{\lambda_n \lambda_k}\, f_n(\mathbf{x}_1) \; .
\end{aligned}
$$

Multiplying once more by $f_l(\mathbf{x}_1)$ and integrating over $\mathbf{D}$, gives

$$
\lambda_k \int_{\mathbf{D}} f_k(\mathbf{x}_1)\, f_l(\mathbf{x}_1)\, d\mathbf{x}_1 = \sum_{n=0}^{\infty} E<\xi_n(\theta)\, \xi_k(\theta)> \sqrt{\lambda_n \lambda_k}\, \delta_{nl}. \quad (2.14)
$$

Then, using equation (2.9) leads to

$$
\lambda_k\, \delta_{kl} = \sqrt{\lambda_k\, \lambda_l} <\xi_k(\theta)\, \xi_l(\theta)> \; . \qquad\qquad (2.15)
$$

Equation (2.15) can be rearranged to give

$$
<\xi_k(\theta)\, \xi_l(\theta)> = \delta_{kl} \; . \qquad\qquad (2.16)
$$

Thus, the random process $w(\mathbf{x}, \theta)$ can be written as

$$
w(\mathbf{x}, \theta) = \bar{w}(\mathbf{x}) + \sum_{n=0}^{\infty} \xi_n(\theta)\, \sqrt{\lambda_n}\, f_n(\mathbf{x}) \; . \qquad\qquad (2.17)
$$

where,

$$
<\xi_n(\theta)> = 0 \quad , \quad <\xi_n(\theta)\, \xi_m(\theta)> = \delta_{nm} \; , \qquad\qquad (2.18)
$$

and λ_n, $f_n(\mathbf{x})$ are solution to equation (2.8). Truncating the series in equation (2.17) at the Mth term, gives

$$w(\mathbf{x},\theta) = \bar{w}(\mathbf{x}) + \sum_{n=0}^{M} \xi_n(\theta) \sqrt{\lambda_n} f_n(\mathbf{x}) . \qquad (2.19)$$

An explicit expression for $\xi_n(\theta)$ can be obtained by multiplying equation (2.11) by $f_n(\mathbf{x})$ and integrating over the domain $\mathbf{D}$. That is,

$$\xi_n(\theta) = \frac{1}{\sqrt{\lambda_n}} \int_{\mathbf{D}} \alpha(\mathbf{x},\theta) f_n(\mathbf{x}) \, d\mathbf{x} . \qquad (2.20)$$

Viewed from a Reproducing Kernel Hilbert Space (RKHS) point of view (Aronszajn, 1950; Parzen, 1959), either of equations (2.11) or (2.20), is an expression for the congruence that maps the Hilbert space spanned by the functions $f_n(\mathbf{x})$ to the Hilbert space spanned by the random process, or equivalently, to the space spanned by the set of random variables $\{\xi_n(\theta)\}$. It is this congruence along with the covariance function of the process that determines uniquely the random process $w(\mathbf{x},\theta)$. Observe the similarity of equations (2.11) and (2.20) with equations (2.7) and (2.8), respectively. Indeed, it can be shown (Parzen, 1959) that if a function f can be represented in terms of linear operations on the family $\{C(\ . \ , \ \mathbf{x}_2 \ ) \ , \ \mathbf{x}_2 \in \ Z\}$, then f belongs to the RKHS corresponding to the kernel $C(\mathbf{x}_1, \mathbf{x}_2)$, and the congruence between the two Hilbert spaces may be expressed in terms of an orthogonal family spanning this RKHS by means of the same linear operations used to represent f in terms of $\{C(.,\mathbf{x}_2) \ , \ \mathbf{x}_2 \in Z\}$. Another point of practical importance is that the expansion given by equation (2.19) can be used in a numerical simulation scheme to obtain numerical realizations of the random process. In fact, this simulation procedure is used in conjunction with the Monte Carlo method for one of the illustrative examples in Chapter IV. It is optimal in the Fourier sense, as it minimizes the mean square error resulting from truncation after a finite number of terms. The expansion is used extensively in the fields of detection, estimation, pattern recognition, and image processing as an efficient tool to store random processes (Devijver and Kittler, 1982). It is worth noting at this point that the Karhunen-Loeve expansion was independently derived in connection with stochastic turbulence problems (Lumley, 1970). In that context, the associated eigenfunctions can be identified with the characteristic eddies of the turbulence field.

It is well known from functional analysis that the steeper a bilinear form decays to zero as a function of one of its arguments, the more terms are needed in its spectral representation in order to reach a preset accuracy. Noting that the Fourier transform operator is a spectral representation, it may be concluded that the faster the autocorrelation function tends to zero, the broader is the corresponding spectral density, and the greater the number of requisite terms to represent the underlying random process by the Karhunen-Loeve expansion.

For the special case of a random process possessing a rational spectrum, the integral eigenvalue problem can be replaced by an equivalent differential equation that is easier to solve (Van Trees, 1968). In the same context, it is reminded that a necessary and sufficient condition for a process to have a finite dimensional Markov realization is that its spectrum be rational (Kree and Soize, 1986). Further, note that analytical solutions for the integral equation (2.8) are obtainable for some quite important and practical forms of the kernel $C(\mathbf{x}_1, \mathbf{x}_2)$ (Juncosa, 1945; Slepian and Pollak, 1961; Van Trees, 1968), some of which are treated later in this chapter.

It may seem that any complete set of functions can be used in-lieu of the eigenfunctions of the covariance kernel in the expansion (2.11). However, it will now be shown that the Karhunen-Loeve expansion as described above, has some desirable properties that make it a preferable choice for some of the objectives of the present approach.

2.3.2 Properties

Error Minimizing Property

The generalized coordinate system defined by the eigenfunctions of the covariance kernel is optimal in the sense that the mean-square error resulting from a finite representation of the process $w(\mathbf{x}, \theta)$ is minimized.

This property can be proved as follows.

Given a complete orthonormal set of functions $h_n(\mathbf{x})$, the process $w(\mathbf{x}, \theta)$ can be approximated in a convergent series of the form

$$w(\mathbf{x}, \theta) = \sum_{n=0}^{\infty} \lambda_n \, \xi_n(\theta) \, h_n(\mathbf{x}) \ . \tag{2.21}$$

Truncating equation (2.21) at the Mth term results in an error ϵ_M equal to

$$\epsilon_M \;=\; \sum_{n=M+1}^{\infty} \lambda_n \; \xi_n(\theta) \; h_n(\mathbf{x}) \; . \tag{2.22}$$

Multiplying equation (2.21) by $h_m(\mathbf{x})$ and integrating throughout gives

$$\xi_m(\theta) \;=\; \frac{1}{\lambda_m} \int_{\mathbf{D}} w(\mathbf{x}, \theta) \; h_m(\mathbf{x}) \; d\mathbf{x} \tag{2.23}$$

where use is made of the orthogonality property of the set $h_n(\mathbf{x})$. Substituting equation (2.23) for $\xi_m(\theta)$ back into equation (2.22), the mean-square error ϵ_M^2 can be written as

$$
\begin{aligned}
\epsilon_M^2 \;=\; & \Bigg[\sum_{m=M+1}^{\infty} \sum_{n=M+1}^{\infty} h_m(\mathbf{x}) \, h_n(\mathbf{x}) \\
& \int_{\mathbf{D}} \int_{\mathbf{D}} <S(\mathbf{x}_1 , \; \theta) \, S(\mathbf{x}_2 , \; \theta)> h_m(\mathbf{x}_1) \, h_n(\mathbf{x}_2) \, d\mathbf{x}_1 \, d\mathbf{x}_2 \Bigg] \\
\;=\; & \sum_{m=M+1}^{\infty} \sum_{n=M+1}^{\infty} h_m(\mathbf{x}) \, h_n(\mathbf{x}) \\
& \int_{\mathbf{D}} \int_{\mathbf{D}} R_{ww}(\mathbf{x}_1, \mathbf{x}_2) \, h_m(\mathbf{x}_1) \, h_n(\mathbf{x}_2) \, d\mathbf{x}_1 \, d\mathbf{x}_2 \; .
\end{aligned}
\tag{2.24}
$$

Integrating equation (2.24) over $\mathbf{D}$ and using the orthonormality of the set $\{h_i(\mathbf{x})\}$ yields

$$\int_{\mathbf{D}} \epsilon_M^2 \, d\mathbf{x} \;=\; \sum_{m=M+1}^{\infty} \int_{\mathbf{D}} \int_{\mathbf{D}} R_{ww}(\mathbf{x}_1, \mathbf{x}_2) \, h_m(\mathbf{x}_1) \, h_m(\mathbf{x}_2) \, d\mathbf{x}_1 \, d\mathbf{x}_2 \; . \tag{2.25}$$

The problem, then, is to minimize $\int_{\mathbf{D}} \epsilon_M^2$ subject to the orthonormality of the functions $h_n(\mathbf{x})$. In other words, the solution minimizes the functional given by the equation

$$
\begin{aligned}
\mathbf{F}[h_k(\mathbf{x})] \;=\; & \sum_{m=M+1}^{\infty} \int_{\mathbf{D}} \int_{\mathbf{D}} R_{ww}(\mathbf{x}_1, \mathbf{x}_2) \, h_m(\mathbf{x}_1) \, h_m(\mathbf{x}_2) \, d\mathbf{x}_1 \, d\mathbf{x}_2 \\
& - \lambda_m \left[\int_{\mathbf{D}} h_m(\mathbf{x}) h_m(\mathbf{x}) \, d\mathbf{x} - 1 \right] \; .
\end{aligned}
\tag{2.26}
$$

Differentiating equation (2.26) with respect to $h_i(\mathbf{x})$ and setting the result equal to zero, gives

$$\frac{\partial \mathbf{F}[h_m(\mathbf{x})]}{\partial h_i(\mathbf{x})} = \int_\mathbf{D} \left[\int_\mathbf{D} R_{ww}(\mathbf{x}_1, \mathbf{x}_2)\, h_i(\mathbf{x}_1)\, d\mathbf{x}_1 \;-\; \lambda_i\, h_i(\mathbf{x}_2) \right]\, d\mathbf{x}_2 = 0 \tag{2.27}$$

which is satisfied when

$$\int_\mathbf{D} R_{ww}(\mathbf{x}_1, \mathbf{x}_2)\, h_i(\mathbf{x}_2)\, d\mathbf{x}_2 \;=\; \lambda_i\, h_i(\mathbf{x}_1) \;\square \tag{2.28}$$

Uniqueness of the Expansion

The random variables appearing in an expansion of the kind given by equation (2.10) are orthonormal if and only if the orthonormal functions $\{f_n(\mathbf{x})\}$ and the constants $\{\lambda_n\}$ are respectively the eigenfunctions and the eigenvalues of the covariance kernel as given by equation (2.8).

The "if" part is an immediate consequence of equation (2.11). To show the "only if" part, equation (2.12) can be used with $<\xi_n(\theta)\xi_m(\theta)> = \delta_{mn}$ to obtain

$$C(\mathbf{x}_1, \mathbf{x}_2) \;=\; \sum_{n=0}^\infty \lambda_n\, f_n(\mathbf{x}_1)\, f_n(\mathbf{x}_2) \;. \tag{2.29}$$

Multiplying both sides by $f_m(\mathbf{x}_2)$ and integrating over $\mathbf{D}$ gives

$$\begin{aligned}
\int_\mathbf{D} C(\mathbf{x}_1, \mathbf{x}_2)\, f_m(\mathbf{x}_2)\, d\mathbf{x}_2 &= \sum_{n=0}^\infty \lambda_n\, f_n(\mathbf{x}_1)\, \delta_{mn} \\
&= \lambda_m\, f_m(\mathbf{x}_1) \;\square
\end{aligned} \tag{2.30}$$

In the context of this last theorem, it is interesting to note that some investigators (e.g. Lawrence, 1987) have used an expansion of the kind given by equation (2.11) with orthogonal random variables and orthogonal deterministic functions that do not satisfy equation (2.8). It is obvious that such an expansion cannot form a basis for the representation of random processes.

Expansion of Gaussian Processes

Let $w(\mathbf{x}, \theta)$ be a Gaussian process with covariance function $C(\mathbf{x}_1, \mathbf{x}_2)$. Then $w(\mathbf{x}, \theta)$ has the Karhunen-Loeve decomposition given by equation (2.17) with the random variables $\xi_i(\theta)$ forming a Gaussian vector. That is, any subset of $\xi_i(\theta)$ is jointly Gaussian. Since these random variables are uncorrelated, their Gaussian property implies their independence. Some important consequences derive from this property. Specifically,

$$<\xi_1(\theta) \, , \, ... \, , \, \xi_{2n+1}(\theta)> \; = \; 0 \qquad (2.31)$$

and

$$<\xi_1(\theta) \, , \, ... \, , \, \xi_{2n}(\theta)> \; = \; \sum \prod \, <\xi_i(\theta) \, \xi_j(\theta)> \, , \qquad (2.32)$$

where the summation extends over all the partitions of the set $\{\xi_i(\theta)\}_{i=1}^{2n}$ into sets of two elements, and the product is over all such sets in a given partition. Furthermore, it can be shown (Loeve, 1977) that for Gaussian processes, the Karhunen-Loeve expansion is almost surely convergent.

Other Properties

In addition to the mean-square error minimizing property, the Karhunen-Loeve expansion has some additional desirable properties. Of these, the minimum representation entropy property is worth mentioning. These properties, however, are of no relevance to the present study and will not be discussed any further. A detailed study of the properties of the Karhunen-Loeve expansion is given by Devijver and Kittler (1982).

2.3.3   Solution of the Integral Equation

Preliminary Remarks

The usefulness of the Karhunen-Loeve expansion hinges on the ability to solve the integral equation of the form

$$\int_{\mathbf{D}} C(\mathbf{x}_1, \mathbf{x}_2) \, f(\mathbf{x}_2) d\mathbf{x}_2 \; = \; \lambda f(\mathbf{x}_1) \, , \qquad (2.33)$$

where $C(\mathbf{x}_1, \mathbf{x}_2)$ is an autocovariance function. Equation (2.33) is a homogeneous Fredholm integral equation of the second kind. The theory underlying this kind of equations has been extensively investigated and is well documented in a number of monographs (Mikhlin, 1957). Being an autocovariance function, the kernel $C(\mathbf{x}_1, \mathbf{x}_2)$ is bounded, symmetric, and positive definite. This fact simplifies the ensuing analysis considerably in that it guarantees a number of properties for the eigenfunctions and the eigenvalues that are solution to equation (2.33). Specifically,

1. The set $f_i(\mathbf{x})$ of eigenfunctions is orthogonal and complete.

2. For each eigenvalue λ_k, there correspond at most a finite number of linearly independent eigenfunctions.

3. There are at most a countably infinite set of eigenvalues.

4. The eigenvalues are all positive real numbers.

5. The kernel $C(\mathbf{x}_1, \mathbf{x}_2)$ admits of the following uniformly convergent expansion

$$C(\mathbf{x}_1, \mathbf{x}_2) = \sum_{k=1}^{\infty} \lambda_k \, f_k(\mathbf{x}_1) \, f_k(\mathbf{x}_2) \, . \tag{2.34}$$

The Karhunen-Loeve expansion of a process was derived based on the preceding analytical properties of its covariance function. These properties are independent of the stochastic nature of the process involved, which allows the expansion to be applied to a wide range of processes including nonstationary and multidimensional processes.

Rational Spectra

Of special interest in engineering applications is the class of one-dimensional random processes that can be realized as the stationary output of a linear filter to white noise excitation. These processes have a spectral density of the form

$$S(\omega) = \frac{N(\omega^2)}{D(\omega^2)} \, , \tag{2.35}$$

where $N(.)$ and $D(.)$ are polynomial operators of order n and d respectively. The interest in this class of processes stems from the fact that a necessary and sufficient condition for a process to be realizable as a finite dimensional Markovian process is that its spectral density function be of the form expressed by equation (2.35) (Kree and Soize, 1986). Loosely speaking, the Markovian property of a process implies that the effect of the infinite past on the present is negligible. That is, the process has a finite memory. For stationary process, equation (2.33) becomes,

$$\int_{\mathbf{D}} C(\mathbf{x}_1 - \mathbf{x}_2)\, f(\mathbf{x}_2)\, d\mathbf{x}_2 \;=\; \lambda\, f(\mathbf{x}_1)\;. \tag{2.36}$$

When the domain $\mathbf{D}$ covers the whole real line, the above equation is equivalent to the Wiener-Hopf integral equation, the solution of which may be found explicitly (Paley and Wiener, 1934; Noble, 1958). The case where $\mathbf{D}$ is finite, however, is more relevant to the context of the monograph, and the solution of the associated integral equation is next detailed. Taking into consideration the one-dimensional form of equation (2.4), and substituting for $S(\omega)$ from equation (2.35), equation (2.36) becomes,

$$\int_{\mathbf{D}} f(x_2) \int_{-\infty}^{\infty} e^{-i\omega\,|x_1-x_2|}\, \frac{N(\omega^2)}{D(\omega^2)}\, d\omega\, dx_2 \;=\; \lambda\, f(x_1)\;. \tag{2.37}$$

Differentiating equation (2.37) twice with respect to x_1 is equivalent to multiplying the integrand by ω^2. Thus, applying the differential operator $D[d^2/dx_1^2]$ to this equation yields

$$
\begin{aligned}
\lambda\, D[\frac{d^2}{dx_1^2}]\, f(x_1) \;&=\; \int_{\mathbf{D}} N[\frac{d^2}{dx_2^2}]\, f(x_2)\, \delta(x_1 - x_2)\, dx_2 \\
&=\; N[\frac{d^2}{dx_1^2}]\, f(x_1)\;,
\end{aligned}
\tag{2.38}
$$

where $\delta(.)$ denotes the dirac delta function. Equation (2.38) can be viewed as a reformulation of equation (2.36) as a homogeneous differential equation. This equation may be solved in terms of the parameter λ and of $2d$ arbitrary constants which are calculated by backsubstituting the resulting solution into equation (2.36). Note, parenthetically, that explicit expressions for the transcendental characteristic equation associated with the differential equation (2.38), for a number of kernel functions, are given in Youla (1957). In the remainder of this section, the preceding treatment of equation (2.36) is applied to the important kernel representing the first order Markovian process. This kernel has been used extensively to model processes in a variety of fields (Yaglom, 1962). Further, it is noted that higher order Markovian kernels may be expressed as linear combinations of first order ones. This important kernel is given by the equation

$$C(x_1, x_2) \;=\; e^{-\,|x_1-x_2|/b}\;, \tag{2.39}$$

where b is a parameter with the same units as x and is often termed the correlation length, since it reflects the rate at which the correlation decays between two points of the process. It is assumed that the process is defined over the one dimensional interval $[-a, a]$. Clearly, $C(x_1, x_2)$ can be made rapidly attenuating versus $|x_1 - x_2|$ by selecting a suitable value of the parameter b. In case the domain $\mathbf{D}$ of the problem is the one-dimensional segment $[-a, a]$, the eigenfunctions and eigenvalues of the covariance function given by equation (2.39) are the solutions to the following integral equation (Van Trees, 1968)

$$\int_{-a}^{+a} e^{-c|x_1 - x_2|} f(x_2) \, dx_2 = \lambda \, f(x_1) , \qquad (2.40)$$

where $c = 1/b$. Equation (2.40) can be written as

$$\int_{-a}^{x} e^{-c(x_1 - x_2)} f(x_2) \, dx_2 + \int_{x}^{a} e^{c(x_1 - x_2)} f(x_2) \, dx_2 = \lambda \, f(x_1) . \quad (2.41)$$

Differentiating equation (2.41) with respect to x_1 and rearranging gives

$$\lambda \, f'(x_1) = -c \int_{-a}^{x} e^{-c(x_1 - x_2)} f(x_2) \, dx_2 + c \int_{x}^{+a} e^{c(x_1 - x_2)} f(x_2) \, dx_2 . \qquad (2.42)$$

Differentiating once more with respect to x_1, the following equation is obtained

$$\lambda \, f''(x) = (-2 \, c + c^2 \, \lambda) \, f(x) . \qquad (2.43)$$

Introducing the new variable

$$\omega^2 = \frac{2 \, c - c^2 \, \lambda}{\lambda} , \qquad (2.44)$$

equation (2.43) becomes

$$f''(x) + \omega^2 \, f(x) = 0 \quad - a \leq x \leq +a . \qquad (2.45)$$

To find the boundary conditions associated with the differential equation

(2.45), equations (2.41) and (2.42) are evaluated at $x = -a$ and $x = +a$. After rearrangement, the boundary conditions become

$$cf(a) + f'(a) = 0 \qquad (2.46)$$

$$c\,f\,(-a) - f'(-a) = 0 \,. \qquad (2.47)$$

Thus, the integral equation given by equation (2.45) is transformed into the ordinary differential equation (2.45) with appended boundary conditions given by equations (2.46) and (2.47). It can be shown that $\omega^2 \geq 0$ is the only range of ω for which equation (2.45) is solvable, the solution being given by the equation,

$$f(x) = a_1 \cos(\omega\,x) + a_2 \sin(\omega\,x)\,. \qquad (2.48)$$

Further, applying the boundary conditions specified by equations (2.46) and (2.47), gives

$$\begin{cases} a_1\,(c - \omega\,\tan(\omega a) + a_2\,(\omega + c\,\tan(\omega a)) = 0 \\[2mm] a_1\,(c - \omega\,\tan(\omega a) - a_2\,(\omega + c\,\tan(\omega a)) = 0. \end{cases} \qquad (2.49)$$

Nontrivial solutions exist only if the determinant of the homogeneous system in equation (2.49) is equal to zero. Setting this determinant equal to zero gives the following transcendental equations

$$\begin{cases} c - \omega\,\tan(\omega a) = 0 \\ \qquad and \\ \omega + c\,\tan(\omega a) = 0\,. \end{cases} \qquad (2.50)$$

Denoting the solution of the second of these equations by ω^*, the resulting eigenfunctions are

$$f_n(x) = \frac{\cos(\omega_n x)}{\sqrt{a + \dfrac{\sin(2\omega_n a)}{2\,\omega_n}}} \qquad (2.51)$$

and

$$f_n^*(x) = \frac{\sin(\omega_n^* x)}{\sqrt{a - \dfrac{\sin(2\omega_n^* a)}{2\,\omega_n^*}}}\,, \qquad (2.52)$$

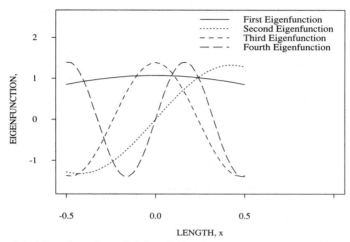

Figure 2.1: Eigenfunctions $f_n(x)$, $-0.5 \leq x \leq 0.5$, $n = 1, 2, 3, 4$; Exponential Covariance, Correlation Length=1.

for even n and odd n respectively. The corresponding eigenvalues are

$$\lambda_n = \frac{2\,c}{\omega_n^2 + c^2} \, . \tag{2.53}$$

and

$$\lambda_n^* = \frac{2\,c}{\omega_n^{*2} + c^2} \, , \tag{2.54}$$

where ω_n and ω_n^* are defined by equation (2.50). Thus, a process $S(x,\theta)$ with covariance function given by equation (2.39) can be expanded as

$$w(x,\theta) = \sum_{n=1}^{\infty} \left[\xi_n \, \sqrt{\lambda_n} \, f_n(x) + \xi_n^* \, \sqrt{\lambda_n^*} \, f_n^*(x) \right] \, . \tag{2.55}$$

Figure (2.1) shows the first 4 eigenfunctions as defined by equations (2.51) and (2.52), for a value of a equal to 0.5 and a value of b equal to 1. Figure (2.2) shows the eigenvalues as given by equation (2.54) for various

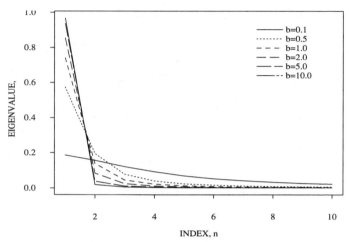

Figure 2.2: Trends of the Eigenvalues λ_n of the Exponential Kernel for Various Values of the Correlation Length b; λ_n assumes values for $n = 1, 2, \ldots$ only.

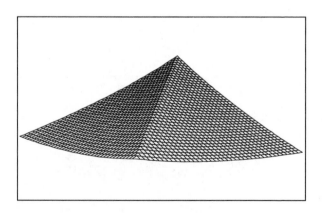

EXPONENTIAL COVARIANCE

Figure 2.3: Exact Covariance Surface versus x_1 and x_2; $|x_1| \leq a$, $|x_2| \leq a$; Correlation Length $= 1$.

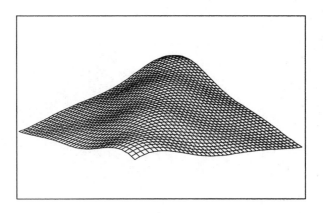

EXPONENTIAL COVARIANCE

Figure 2.4: 4-Term Approximation of Covariance Surface versus x_1 and x_2; $|x_1| \leq a$, $|x_2| \leq a$; Correlation Length = 1.

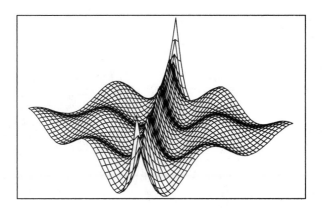

EXPONENTIAL COVARIANCE

Figure 2.5: 4-Term Relative Error Surface of Covariance Approximation versus x_1 and x_2; $|x_1| \leq a$, $|x_2| \leq a$; Maximum Error = 0.1126; Correlation Length = 1.

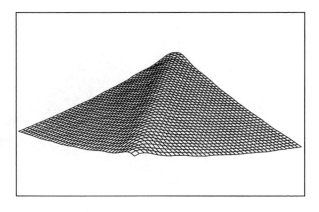

EXPONENTIAL COVARIANCE

Figure 2.6: 10-Term Approximation Covariance Surface versus x_1 and x_2; $|x_1| \leq a$, $|x_2| \leq a$; Correlation Length = 1.

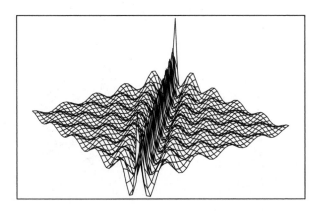

EXPONENTIAL COVARIANCE

Figure 2.7: 10-Term Relative Error Surface of Covariance Approximation versus x_1 and x_2; $|x_1| \leq a$, $|x_2| \leq a$; Maximum Error = 0.0425; Correlation Length = 1.

values of b. Note that the smaller the value of b, the more contribution should be expected from terms associated with smaller eigenvalues. Figures (2.3)-(2.7) show the exact kernel, its four term approximation, its ten term approximation and the associated errors, respectively.

Nonrational Spectra

There is no general method for the solution of the integral equation (2.33) corresponding to nonrational spectra. Several of these equations have been investigated and explicit solutions have been obtained for certain covariance kernels. The method described in the previous section may be applied successfully to a number of these kernels as will be demonstrated for the case of the triangular kernel given by the equation

$$C(x_1, x_2) = 1 - d\,|x_1 - x_2|\,, \quad |x_1 - x_2| \in [0, \frac{1}{d}]\,. \tag{2.56}$$

Here, d is a parameter which can be used to adjust the distance $|x_1 - x_2|$ of null correlation between $w(x_1\,,\,\theta)$ and $w(x_2\,,\,\theta)$. Consider realizations of this process on the interval $[0, a]$. The eigenfunctions and eigenvalues of $C(x_1, x_2)$ are obtained as the solution to the integral equation

$$\int_0^a (1 - d\,|x_1 - x_2|)\,f_n(x_2)\,dx_2 = \lambda_n\,f_n(x_1)\,. \tag{2.57}$$

Differentiating equation (2.57) twice with respect to x_2, the following equivalent differential equation is obtained

$$f_n''(x) + \omega_n^2\,f_n(x) = 0\,, \quad 0 \le x \le a\,. \tag{2.58}$$

The associated boundary conditions are given by the equations

$$f'_n(a) = -f'_n(0)\,, \tag{2.59}$$

$$f'_n(0) = \frac{f_n(0) + f_n(a)}{\dfrac{2}{d} - a}\,, \tag{2.60}$$

and

$$\omega_n = \sqrt{\frac{2d}{\lambda_n}} \, . \tag{2.61}$$

The solution of equation (2.58) subjected to the boundary conditions described by equations (2.59) and (2.60) is, for n odd,

$$f_n(x) = \frac{\cos(\omega_n x) + \tan(\frac{\omega_n a}{2})\sin(\omega_n x)}{\sqrt{a + (\tan^2(\frac{\omega_n a}{2}) - 1)(\frac{a}{2} - \frac{\sin(2\omega_n a)}{4\omega_n}) + \frac{\sin^2(\omega_n a)}{\omega_n}\tan(\frac{\omega_n a}{2})}} \tag{2.62}$$

and, for n even,

$$f_n(x) = \frac{\cos(\omega_n x)}{\sqrt{\frac{a}{2} + \frac{\sin(2\omega_n a)}{2\omega_n}}} \, , \tag{2.63}$$

where ω_n is the solution to the following equation

$$\tan(\frac{\omega_n a}{2}) = \frac{2}{\omega_n \left(\frac{2}{d} - a\right)} \, , \quad for \; n = 2, 4, 6... \tag{2.64}$$

and

$$\omega_n = n\frac{\pi}{a} \, , \quad for \; n = 1, 3, 5... \tag{2.65}$$

Figure (2.8) shows a plot of the first four eigenfunctions associated with this kernel. The eigenvalues are shown in Figure (2.9) for various values of d. Figures (2.10)-(2.14) show the exact kernel, its four-term approximation, its ten-term approximation and the corresponding errors.

Another kernel that may be treated by the same method is the kernel of the Wiener process. It is given by the equation

$$C(x_1, x_2) = \min(x_1, x_2), \quad (x_1, x_2) \in [0, T] \times [0, T] \, . \tag{2.66}$$

The resulting normalized eigenfunctions and eigenvalues are

$$f_n(x_1) = \sqrt{2} \, \sin(\frac{x_1}{\sqrt{\lambda_n}}) \, , \tag{2.67}$$

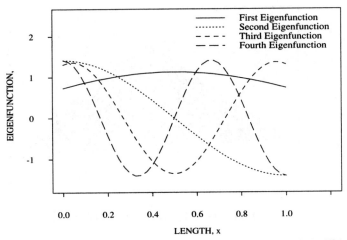

Figure 2.8: Eigenfunctions $f_n(x)$, $0 \leq x \leq 1$, $n = 1, 2, 3, 4$; Triangular Covariance, Correlation Length=1.

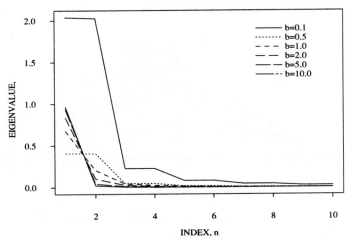

Figure 2.9: Trends of the Eigenvalues λ_n of the Triangular Kernel for Various Values of the Correlation Length b; λ_n assumes values for $n = 1, 2, \dots$ only.

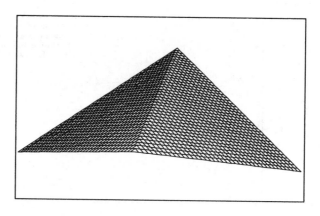

TRIANGULAR COVARIANCE

Figure 2.10: Exact Covariance Surface versus x_1 and x_2; $0 \leq x_1 \leq a$, $0 \leq x_2 \leq a$; Correlation Length = 1.

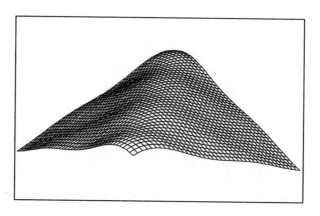

TRIANGULAR COVARIANCE

Figure 2.11: 4-Term Approximation Covariance Surface versus x_1 and x_2; $0 \leq x_1 \leq a$, $0 \leq x_2 \leq a$; Correlation Length = 1.

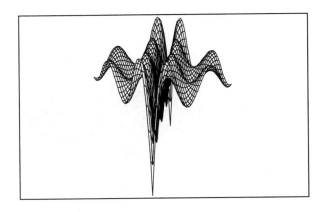

TRIANGULAR COVARIANCE

Figure 2.12: 4-Term Relative Error Surface of Covariance Approximation versus x_1 and x_2; $0 \leq x_1 \leq a$, $0 \leq x_2 \leq a$; Maximum Error = 0.1226; Correlation Length = 1.

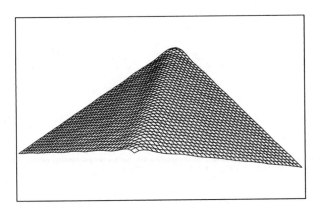

TRIANGULAR COVARIANCE

Figure 2.13: 10-Term Approximation Covariance Surface versus x_1 and x_2; $0 \leq x_1 \leq a$, $0 \leq x_2 \leq a$; Correlation Length = 1.

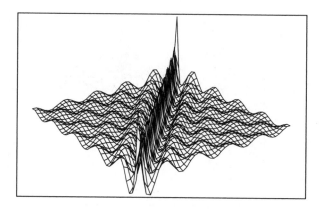

TRIANGULAR COVARIANCE

Figure 2.14: 10-Term Relative Error Surface of Covariance Approximation versus x_1 and x_2; $0 \leq x_1 \leq a$, $0 \leq x_2 \leq a$; Maximum Error $= 0.0425$; Correlation Length $= 1$.

and

$$\lambda_n = \frac{4T^2}{(\pi^2(2n+1)^2)} \ , \ n = 0, 1, \ldots \tag{2.68}$$

In addition to the examples discussed above, the same method may be applied to kernels that are integrals of the triangular kernel, or kernels that are sums of exponential and triangular kernels (Kailath, 1966). Note that the Wiener process is an example of a nonstationary process, a fact that emphasizes the generality of the Karhunen-Loeve expansion and its applicability to such processes.

Another process that has attracted much attention in the literature has a covariance function given by the function

$$C(x_1, x_2) = e^{-(x_1+x_2)} \ e^{-|x_1-x_2|/b} \ . \tag{2.69}$$

This process is a uniformly modulated nonstationary process (Liu, 1970), with modulating function $\sqrt{e^{-x}}$; it is useful in modeling processes whose correlation function has maximum value decaying with time. The eigenfunctions and eigenvalues of the covariance kernel given by equation (2.69)

can be related to those of the nonsymmetric kernel given by the following
equation (Juncosa, 1945),

$$C(x_1, x_2) = e^{-2x_1} e^{-|x_1 - x_2|/b} . \qquad (2.70)$$

Differentiating equation (2.33) twice with respect to x_1, after substituting for
$C(x_1, x_2)$ from equation (2.70), leads to a homogeneous differential equation
of the Bessel type. The final solution (Juncosa, 1945) is

$$f_n(x_1) = J_{1/b} \left[\frac{2 e^{-x}}{\sqrt{b \lambda_n}} \right] \qquad (2.71)$$

where $J_k[.]$ is the Bessel function of order k and

$$\lambda_n = \frac{4}{b r_n^2} \; ; \; J_{1/b-1}[r_n] = 0 \; ; \; r_n > 0 \; , \; n = 1 \, , \, 2 \, , \, ... \qquad (2.72)$$

Another process that is useful in engineering applications has a covariance kernel given by the equation

$$C(x_1, x_2) = \frac{\sin(\Omega(x_1 - x_2))}{\pi(x_1 - x_2)} , \qquad (2.73)$$

where Ω is constant. This is a truncated white noise process, whose power
spectral density vanishes outside a certain interval where it is constant. The
associated eigenfunctions and eigenvalues were obtained in explicit form by
Slepian and Pollak (1961). They are the angular prolate spheroidal wave
functions and the radial prolate spheroidal wave functions respectively. They
have some quite interesting properties that are, however, beyond the scope
of the present work (Slepian, 1964, 1965, 1968, 1977; Landau et.al, 1961,
1962).

A final remark is in order concerning the choice of the domain **D** of
definition of the random process being investigated. Taking **D** to be the
finite domain over which the process is being observed may often be the most
obvious choice. Clearly, this choice does not induce the ergodic assumption
(Lin, 1967) for the process, which involves observing infinite length records.
This is by no means a handicap of this approach since the ergodic assumption
is usually introduced for convenience and is not necessary for the present
study. If ergodicity is needed for some particular problem, then it may be
recovered by extending the limits of integration in equation (2.33) to infinity.

This modification may be convenient numerically if, for instance, an explicit solution of the integral equation is available for the infinite domain and not for the finite one.

Numerical Solution

In this section, a Galerkin type procedure is described for the solution of the Fredholm equation (2.33). In Chapter IV, this procedure will be illustrated through its application to a curved geometry two-dimensional problem. Let $h_i(\mathbf{x})$ be a complete set of functions in the Hilbert space $\mathbf{H}$. Each eigenfunction of the kernel $C(\mathbf{x}_1, \mathbf{x}_2)$ may be represented as

$$f_k(\mathbf{x}) = \sum_{i=1}^{N} d_i^{(k)} \, h_i(\mathbf{x}) \, , \tag{2.74}$$

with an error ϵ_N resulting from truncating the summation after the Nth term. This error is equal to the difference between the left hand side and the right hand side of equation (2.33). Substituting equation (2.74) into equation (2.33) yields the following expression for the error

$$\epsilon_N = \sum_{i=1}^{N} d_i^{(k)} \left[\int_{\mathbf{D}} C(\mathbf{x}_1, \mathbf{x}_2) \, h_i(\mathbf{x}_2) \, d\mathbf{x}_2 \, - \, \lambda_n \, h_i(\mathbf{x}_1) \right] . \tag{2.75}$$

Requiring the error to be orthogonal to the approximating space yields equations of the following form,

$$(\epsilon_N \, , \, h_j(\mathbf{x})) = 0 \, , \quad j = 1 \, ,..., \, N \, . \tag{2.76}$$

Equivalently,

$$\sum_{i=1}^{N} d_i^{(k)} \left[\int_{\mathbf{D}} \left[\int_{\mathbf{D}} C(\mathbf{x}_1, \mathbf{x}_2) \, h_i(\mathbf{x}_2) \, d\mathbf{x}_2 \right] h_j(\mathbf{x}_1) \, d\mathbf{x}_1 \right.$$

$$\left. - \lambda_n \int_{\mathbf{D}} h_i(\mathbf{x}) \, h_j(\mathbf{x}) \, d\mathbf{x}_1 \right] = 0 \, . \tag{2.77}$$

Denoting

$$\mathbf{C}_{ij} = \int_{\mathbf{D}} \int_{\mathbf{D}} C(\mathbf{x}_1, \mathbf{x}_2) \, h_i(\mathbf{x}_2) \, d\mathbf{x}_2 \, h_j(\mathbf{x}_1) \, d\mathbf{x}_1 \, d\mathbf{x}_2 \, , \tag{2.78}$$

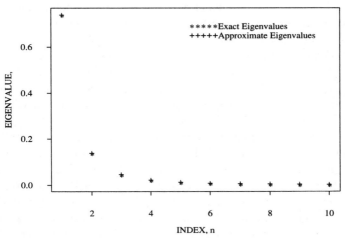

Figure 2.15: Eigenvalues λ_n, of the Exponential Covariance Kernel, $n = 1, ..., 9$; Exact and Approximate values.

$$\mathbf{B}_{ij} = \int_{\mathbf{D}} h_i(\mathbf{x}) \, h_j(\mathbf{x}) \, d\mathbf{x} \, , \tag{2.79}$$

$$\mathbf{D}_{ij} = d_i^{(j)} \, , \tag{2.80}$$

and

$$\mathbf{\Lambda}_{ij} = \delta_{ij}\lambda_i \, , \tag{2.81}$$

equation (2.77) becomes

$$\mathbf{C} \, \mathbf{D} = \mathbf{\Lambda} \, \mathbf{B} \, \mathbf{D} \, , \tag{2.82}$$

where $\mathbf{C}$, $\mathbf{B}$ and $\mathbf{D}$ are three N-dimensional matrices whose elements are given by equations (2.78)-(2.79). Equation (2.82) represents a generalized algebraic eigenvalue problem which may be solved for the matrix $\mathbf{D}$ and the eigenvalues λ_k. Backsubstituting into equation (2.74) yields the eigenfunctions of the covariance kernel. The preceding procedure can be implemented using piecewise polynomials as the basis for the expansion. With this choice of basis functions, the columns of the matrix $\mathbf{D}$ become the eigenvectors computed at the respective nodal points of the induced mesh, and the ij^{th} element of the matrix $\mathbf{C}$ becomes the weighted correlation between the process at nodes i and j. Note that both matrices $\mathbf{C}$ and $\mathbf{B}$ are symmetric positive definite, a fact that substantially simplifies the numerical solution. Figure (2.15) shows the exact eigenvalues of the kernel given by equation (2.39) and the results from the algorithm described above; note the excellent agreement. Obviously, the eigenvectors and eigenvalues computed based on the above scheme provide convergent estimates to the true values. Certain properties of these particular estimates make this scheme computationally attractive. Specifically, the Galerkin scheme described above can be shown to be equivalent to a variational treatment of the problem. This property ensures that the computed eigenvalues are a lower bound of the correspondingly numbered exact eigenvalues. This implies that the convergence of each eigenvalue is monotonic in N. Further, note that the accuracy in estimating the eigenvalues is better than that achieved for the eigenfunctions (Delves and Mohamed, 1985) .

2.4   Homogeneous Chaos

2.4.1   Preliminary Remarks

It is clear from the preceding discussion that the implementation of the Karhunen-Loeve expansion requires knowing the covariance function of the process being expanded. As far as the system under consideration is concerned, this implies that the expansion can be used for the random coefficients in the operator equation. However, it cannot be implemented for the solution process, since its covariance function and therefore the corresponding eigenfunctions are not known. An alternative expansion is clearly needed which circumvents this problem. Such an expansion could involve a basis of known random functions with deterministic coefficients to be found by minimizing some norm of the error resulting from a finite representation. This should be construed as similar to the Fourier series solution of deterministic differential equations, whereby the series coefficients are determined so as to satisfy some optimality criterion. To clarify this important idea further, a useful modification of the problem suggested by equation (1.13) is noted. For this, equation (1.13) is rewritten as

$$u = h\,[\xi_i(\theta), x] \tag{2.83}$$

where $h[.]$ is a nonlinear functional of its arguments. In equation (2.83), the random processes involved have all been replaced by their corresponding Karhunen-Loeve representations. It is clear now that what is required is a nonlinear expansion of $h[.]$ in terms of the set of random variables $\xi_i(\theta)$. If the processes defining the operator are Gaussian, this set is a sampled derivative of the Wiener process (Doob, 1953). In this case, equation (2.83) involves functionals of the Brownian motion. This is exactly what the concept of Homogeneous Chaos provides. This concept was first introduced by Wiener (1938) and consists of an extension of the then obscure Volterra's work on the generalization of Taylor series to functionals (Volterra, 1913). Wiener's contributions were the result of his investigations of nonlinear functionals of the Brownian motion. Based on Wiener's ideas, Cameron and Martin (1947) constructed an orthogonal basis for nonlinear functionals in terms of Fourier-Hermite functionals. Wiener's Homogeneous Chaos was subsequently refined by Itô (1951) into what is known as the "Multiple Wiener Integral". About the same time that this analytical and measure-theoretic development of the theory was being pursued, Murray and Von-Neumann (1936) were establishing the parallel algebraic structure of rings of operators. Wiener's theory was further developed through research efforts that led to a series of reports (Bose, 1956; George, 1959). Numerical implementation of the basic ideas as well as convergence properties were addressed in these reports. This theory has drawn the interest of investigators in the fields of communication (Yasui, 1979), neuro-science (Palm and Poggio, 1977), engineering mechanics (Jahedi and Ahmadi, 1983), statistical physics (Imamura et.al, 1965a-b) and mathematics (Hida and Ikeda, 1965; McKean, 1973; Huang and Cambanis, 1978, 1979; Kallianpur, 1980; Engels, 1982). In particular, Yasui (1979), Engels (1982) and Kallianpur (1980) have attempted to develop a unified treatment of the Volterra series, the Wiener series, the Cameron-Martin expansion and the Itô approach. They have concluded that the last three approaches are equivalent and that they are superior to the Volterra series in terms of their convergence properties and their applicability. In the same manner that the Homogeneous Chaos can be viewed as an orthogonal development for nonlinear functionals with Gaussian measure, the Discrete Chaos (Wintner and Wiener, 1943; Hida and Ikeda, 1965; Ogura, 1972) is an orthogonal development with respect to the Poisson measure. Extensions to general measures have been investigated by Segall and Kailath (1976).

2.4.2 Definitions and Properties

Let $\{\xi_i(\theta)\}_{i=1}^{\infty}$ be a set of orthonormal Gaussian random variables. Consider the space $\hat{\Gamma}_p$ of all polynomials in $\{\xi_i(\theta)\}_{i=1}^{\infty}$ of degree not exceeding p. Let Γ_p represent the set of all polynomials in $\hat{\Gamma}_p$ orthogonal to $\hat{\Gamma}_{p-1}$. Finally, let $\bar{\Gamma}_p$ be the space spanned by Γ_p. Then, the subspace $\bar{\Gamma}_p$ of Θ is called the pth Homogeneous Chaos, and Γ_p is called the Polynomial Chaos of order p.

Based on the above definitions, the Polynomial Chaoses of any order p consist of all orthogonal polynomials of order p involving any combination of the random variables $\{\xi_i(\theta)\}_{i=1}^{\infty}$. It is clear, then, that the number of Polynomial Chaoses of order p, which involve a specific random variable out of the set $\{\xi_i(\theta)\}_{i=1}^{\infty}$ increases with p. This fact plays an important role in connection with the finite dimensional Polynomial Chaoses to be introduced in the sequel. Furthermore, since random variables are themselves functions, it becomes clear that Polynomial Chaoses are functions of functions and are therefore functionals.

The set of Polynomial Chaoses is a linear subspace of the space of square-integrable random variables Θ, and is a ring with respect to the functional multiplication $\Gamma_p\Gamma_l(\omega) = \Gamma_p(\omega)\Gamma_l(\omega)$. In this context, square integrability must be construed to be with respect to the probability measure defining the random variables. Denoting the Hilbert space spanned by the set $\{\xi_i(\theta)\}$ by $\Theta(\xi)$, the resulting ring is denoted by $\Phi_{\Theta(\xi)}$, and is called the ring of functions generated by $\Theta(\xi)$. Then, it can be shown that under some general conditions, the ring $\Phi_{\Theta(\xi)}$ is dense in the space Θ (Kakutani, 1961). This means that any square-integrable random function $(\Omega \to R)$ can be approximated as closely as desired by elements from $\Phi_{\Theta(\xi)}$. Thus, any element $\mu(\theta)$ from the space Θ admits the following representation,

$$\mu(\theta) = \sum_{p \geq 0} \sum_{n_1+\ldots+n_r=p} \sum_{\rho_1,\ldots,\rho_r} a_{\rho_1\ldots\rho_r}^{n_1\ldots n_r} \, \Gamma_p(\xi_{\rho_1}(\theta), \ldots, \xi_{\rho_r}(\theta)) \,, \qquad (2.84)$$

where $\Gamma_p(.)$ is the Polynomial Chaos of order p. The superscript n_i refers to the number of occurrences of $\xi_{\rho_i}(\theta)$ in the argument list for $\Gamma_p(.)$. Also, the double subscript provides for the possibility of repeated arguments in the argument list of the Polynomial Chaoses, thus preserving the generality of the representation given by equation (2.84). Briefly stated, the Polynomial Chaos appearing in equation (2.84) involves r distinct random variables out of the set $\{\xi_i(\theta)\}_{i=1}^{\infty}$, with the kth random variable $\xi_k(\theta)$ having multiplicity n_k, and such that the total number of random variables involved is equal to the order p of the Polynomial Chaos. The Polynomial Chaoses of any order will be assumed to be symmetrical with respect to their arguments. Such a symmetrization is always possible. Indeed, a symmetrical polynomial can be obtained from a non-symmetrical one by taking the average of the polynomial over all permutations of its arguments. The form of the coefficients appearing in equation (2.84) can then be simplified, resulting in the following expanded expression for the representation of random variables,

$$\mu(\theta) = a_0 \, \Gamma_0 + \sum_{i_1=1}^{\infty} a_{i_1} \Gamma_1(\xi_{i_1}(\theta)) \tag{2.85}$$

$$+ \sum_{i_1=1}^{\infty} \sum_{i_2=1}^{i_1} a_{i_1 i_2} \Gamma_2(\xi_{i_1}(\theta), \xi_{i_2}(\theta))$$

$$+ \sum_{i_1=1}^{\infty} \sum_{i_2=1}^{i_1} \sum_{i_3=1}^{i_2} a_{i_1 i_2 i_3} \Gamma_3(\xi_{i_1}(\theta), \xi_{i_2}(\theta), \xi_{i_3}(\theta))$$

$$+ \sum_{i_1=1}^{\infty} \sum_{i_2=1}^{i_1} \sum_{i_3=1}^{i_2} \sum_{i_4=1}^{i_3} a_{i_1 i_2 i_3 i_4} \Gamma_4(\xi_{i_1}(\theta), \xi_{i_2}(\theta), \xi_{i_3}(\theta), \xi_{i_4}(\theta)) + \ldots,$$

where $\Gamma_p(.)$ are successive Polynomial Chaoses of their arguments, the expansion being convergent in the mean-square sense. The upper limits on the summations in equation (2.85) reflect the symmtery of the Polynomial Chaoses with respect to their arguments, as discussed above. The Polynomial Chaoses of order greater than one have mean zero. Polynomials of different order are orthogonal to each other; so are same order polynomials with different argument list. At times in the ensuing developments, it will prove notationally expedient to rewrite equation (2.85) in the form

$$\mu(\theta) = \sum_{j=0}^{\infty} \hat{a}_j \, \Psi_j[\boldsymbol{\xi}(\theta)], \tag{2.86}$$

where there is a one-to-one correspondence between the functionals $\Psi[.]$ and $\Gamma[.]$, and also between the coefficients $\hat{a}_j$ and $a_{i_1 \ldots i_r}$ appearing in equation (2.85). Implicit in equation (2.85) is the assumption that the expansion

(2.85) is carried out in the order indicated by that equation. In other words, the contribution of polynomials of lower order is accounted for first.

Up to now, and throughout the previous theoretical development, the symbol θ has been used to emphasize the random character of the quantities involved. It is felt that, although somewhat cumbersome, this notation underlines the fact that a random variable is a *function* defined over the space of events of which θ is an element. Having noted this, the symbol θ will be deleted in the ensuing development whenever the random nature of a certain quantity is obvious from the context.

As defined above, each Polynomial Chaos is a function of the infinite set $\{\xi_i\}$, and is therefore an infinite dimensional polynomial. In a computational setting, however, this infinite set has to be replaced by a finite one. In view of that, it seems logical to introduce the concept of a finite dimensional Polynomial Chaos. Specifically, the n-dimensional Polynomial Chaos of order p is the subset of the Polynomial Chaos of order p, as defined above, which is a function of only n of the uncorrelated random variables ξ_i. As $n \to \infty$, the Polynomial Chaos as defined previously is recovered. Obviously, the convergence properties of a representation based on the n-dimensional Polynomial Chaoses depend on n as well as on the choice of the subset $\{\xi_{\lambda_i}\}_{i=1}^n$ out of the infinite set. In the ensuing analysis, this choice will be based on the Karhunen-Loeve expansion of an appropriate random process. Since the finite dimensional Polynomial Chaos is a subset of the (infinite-dimensional) Polynomial Chaos, the same symbol will be used for both, with the dimension being specified. Note that for this case, the infinite upper limit on the summations in equation (2.85) is replaced by a number equal to the dimension of the Polynomials involved. For clarity, the two-dimensional counterpart of equation (2.85) is rewritten, in a fully expanded form, as

$$
\begin{aligned}
\mu(\theta) = \ & a_0\, \Gamma_0 \ + \ a_1\, \Gamma_1(\xi_1) \ + \ a_2\, \Gamma_1(\xi_2) \hspace{2cm} (2.87) \\
& + \ a_{11}\, \Gamma_2(\xi_1,\xi_1) \ + \ a_{12}\Gamma_2(\xi_2,\xi_1) \ + \ a_{22}\Gamma_2(\xi_2,\xi_2) \\
& + \ a_{111}\, \Gamma_3(\xi_1,\xi_1,\xi_1) \ + \ a_{211}\, \Gamma_3(\xi_2,\xi_1,\xi_1) \ + \ a_{221}\, \Gamma_3(\xi_2,\xi_2,\xi_1) \\
& + \ a_{222}\, \Gamma_3(\xi_2,\xi_2,\xi_2) \ \ldots \ .
\end{aligned}
$$

In view of this last equation, it becomes clear that, except for a different indexing convention, the functionals $\Psi[.]$ and $\Gamma[.]$ are identical. In this regard, equation (2.87) can be recast in terms of $\Psi_j[.]$ as follows

$$
\mu(\theta) = \ \hat{a}_0\Psi_0 \ + \ \hat{a}_2\Psi_2 \ + \ \hat{a}_3\Psi_3 \ + \ \hat{a}_4\Psi_4 \ + \ \hat{a}_5\Psi_5
$$

$$+ \hat{a}_6\Psi_6 + \hat{a}_7\Psi_7 + \hat{a}_8\Psi_8 + \hat{a}_9\Psi_9 + \ldots , \qquad (2.88)$$

from which the correspondence between $\Psi[.]$ and $\Gamma[.]$ is evident. For example, the term $a_{211}\Gamma_3(\xi_2, \xi_1, \xi_1)$ of equation (2.87) is identified with the term $\hat{a}_7\Psi_7$ of equation (2.88).

2.4.3 Construction of the Polynomial Chaos

A direct approach to construct the successive Polynomial Chaoses is to start with the set of homogeneous polynomials in $\{\xi_i(\theta)\}$ and to proceed, through a sequence of orthogonalization procedures. The zeroth order polynomial is a constant and it can be chosen to be 1. That is

$$\Gamma_0 = 1 . \qquad (2.89)$$

The first order polynomial has to be chosen so that it is orthogonal to all zeroth order polynomials. In this context, orthogonality is understood to be with respect to the inner-product defined by equation (1.8). Since the set $\{\xi_i\}$ consists of zero-mean elements, the orthogonality condition implies

$$\Gamma_1(\xi_i) = \xi_i . \qquad (2.90)$$

The second order Polynomial Chaos consists of second order polynomials in $\{\xi_i\}$ that are orthogonal to both constants and first order polynomials. Formally, a second order polynomial can be written as

$$\Gamma_2(\xi_{i_1}, \xi_{i_2}) = a_0 + a_{i_1}\xi_{i_1} + a_{i_2}\xi_{i_2} + a_{i_1 i_2}\xi_{i_1}\xi_{i_2} , \qquad (2.91)$$

where the constants are so chosen as to satisfy the orthogonality conditions. The second of these requires that

$$<\Gamma_2(\xi_{i_1}, \xi_{i_2})\, \xi_{i_3}> = 0 . \qquad (2.92)$$

This results in the following equation

$$a_{i_1}\delta_{i_1 i_3} + a_{i_2}\delta_{i_2 i_3} = 0 . \qquad (2.93)$$

Allowing i_3 to be equal to i_1 and i_2 successively, permits the evaluation of the coefficients a_{i_1} and a_{i_2} as

$$a_{i_1} = 0 , \quad a_{i_2} = 0 . \qquad (2.94)$$

The first orthogonality condition results in

$$a_0 + a_{i_1 i_2} \delta_{i_1 i_2} = 0 . \tag{2.95}$$

Equation (2.95) can be normalized by requiring that

$$a_{i_1 i_2} = 1 . \tag{2.96}$$

This leads to

$$a_0 = - \delta_{i_1 i_2} . \tag{2.97}$$

Thus, the second Polynomial Chaos can be expressed as

$$\Gamma_2(\xi_{i_1}, \xi_{i_2}) = \xi_{i_1} \xi_{i_2} - \delta_{i_1 i_2} . \tag{2.98}$$

In a similar manner, the third order Polynomial Chaos has the general form

$$
\begin{aligned}
\Gamma_3(\xi_{i_1}, \xi_{i_2}, \xi_{i_3}) =& a_0 + a_{i_1}\xi_{i_1} + a_{i_2}\xi_{i_2} + a_{i_3}\xi_{i_3} + a_{i_1 i_2}\xi_{i_1}\xi_{i_2} \\
&+ a_{i_1 i_3}\xi_{i_1}\xi_{i_3} + a_{i_2 i_3}\xi_{i_2}\xi_{i_3} + a_{i_1 i_2 i_3}\xi_{i_1}\xi_{i_2}\xi_{i_3},
\end{aligned} \tag{2.99}
$$

with conditions of being orthogonal to all constants, first order polynomials, and second order polynomials. The first of these conditions implies that

$$<\Gamma_3(\xi_{i_1}, \xi_{i_2}, \xi_{i_3})> = 0 . \tag{2.100}$$

That is,

$$a_0 + a_{i_1 i_2} \delta_{i_1 i_2} + a_{i_1 i_3} \delta_{i_1 i_3} + a_{i_2 i_3} \delta_{i_2 i_3} = 0 . \tag{2.101}$$

The second condition implies that

$$<\Gamma_3(\xi_{i_1}, \xi_{i_2}, \xi_{i_3})\, \xi_{i_4}> = 0 , \tag{2.102}$$

which leads to

$$a_{i_1}\delta_{i_1 i_4} + a_{i_2}\delta_{i_2 i_4} + a_{i_3}\delta_{i_3 i_4} + a_{i_1 i_2 i_3} <\xi_{i_1}\, \xi_{i_2}\, \xi_{i_3}\, \xi_{i_4}> . \tag{2.103}$$

The last orthogonality condition is equivalent to

$$<\Gamma_3(\xi_{i_1}, \xi_{i_2}, \xi_{i_3})\, \xi_{i_4}\, \xi_{i_5}> = 0 , \tag{2.104}$$

which results in

$$a_0 \, \delta_{i_4 i_5} \, a_{i_1 i_2} {<}\xi_{i_1}\xi_{i_2}\xi_{i_4}\xi_{i_5}{>} \; + \; a_{i_1 i_3}{<}\xi_{i_1}\xi_{i_3}\xi_{i_4}\xi_{i_5}{>}$$
$$+ \; a_{i_2 i_3}{<}\xi_{i_2}\xi_{i_3}\xi_{i_4}\xi_{i_5}{>} \;\; = 0. \qquad (2.105)$$

The above equations can be normalized by requiring that

$$a_{i_1 i_2 i_3} \; = \; 1 \, . \qquad (2.106)$$

Then equation (2.103) becomes

$$a_{i_1}\delta_{i_1 i_4} \; + \; a_{i_2}\delta_{i_2 i_4} \; + \; a_{i_3}\delta_{i_3 i_4} \; + \; {<}\xi_{i_1}\xi_{i_2}\xi_{i_3}\xi_{i_4}{>} \; = \; 0 \qquad (2.107)$$

Due to the Gaussian property of the set $\{\xi_i\}$, the following equation holds

$${<}\xi_{i_1}\xi_{i_2}\xi_{i_3}\xi_{i_4}{>} \; = \; \delta_{i_1 i_2}\delta_{i_3 i_4} \; + \; \delta_{i_1 i_3}\delta_{i_2 i_4} \; + \; \delta_{i_1 i_4}\delta_{i_2 i_3} \, . \qquad (2.108)$$

Substituting for the expectations in equations (2.107) and (2.105) yields

$$a_{i_1}\delta_{i_1 i_4} \; + \; a_{i_2}\delta_{i_2 i_4} \; + \; a_{i_3}\delta_{i_3 i_4}$$
$$+ \; \delta_{i_1 i_2}\delta_{i_3 i_4} \; + \; \delta_{i_1 i_3} \, \delta_{i_2 i_4} \; + \; \delta_{i_1 i_4}\delta_{i_2 i_3} \; = \; 0 \, , \qquad (2.109)$$

and

$$a_0 \, \delta_{i_4 i_5} \; + \; a_{i_1 i_2} \, [\; \delta_{i_1 i_2}\delta_{i_4 i_5} \; + \; \delta_{i_1 i_4}\delta_{i_2 i_5} \; + \; \delta_{i_1 i_5}\delta_{i_2 i_4} \;]$$
$$+ \; a_{i_1 i_3} \, [\; \delta_{i_1 i_3}\delta_{i_4 i_5} \; + \; \delta_{i_1 i_4}\delta_{i_3 i_5} \; + \; \delta_{i_1 i_5}\delta_{i_3 i_4} \;]$$
$$+ \; a_{i_2 i_3} \, [\; \delta_{i_2 i_3}\delta_{i_4 i_5} \; + \; \delta_{i_2 i_4}\delta_{i_3 i_5} \; + \; \delta_{i_2 i_5}\delta_{i_3 i_4} \;] \; = \; 0. \qquad (2.110)$$

Substituting for a_0 from equation (2.101), equation (2.110) can be rewritten as

$$a_{i_1 i_2} \, [\; \delta_{i_1 i_4}\delta_{i_2 i_5} \; + \delta_{i_1 i_5}\delta_{i_2 i_4} \;] \; + \; a_{i_1 i_3} \, [\; \delta_{i_1 i_4}\delta_{i_3 i_5} \; + \; \delta_{i_1 i_5}\delta_{i_3 i_4} \;]$$
$$+ a_{i_2 i_3} \, [\; \delta_{i_2 i_4}\delta_{i_3 i_5} \; + \; \delta_{i_2 i_5}\delta_{i_3 i_4} \;] \; = \; 0 \, . \qquad (2.111)$$

From equation (2.111), the coefficients $a_{i_1 i_2}$, $a_{i_1 i_3}$, and $a_{i_2 i_3}$ can be evaluated as

$$a_{i_1 i_2} \; = \; 0$$
$$a_{i_1 i_3} \; = \; 0 \qquad (2.112)$$
$$a_{i_2 i_3} \; = \; 0 \, .$$

Using equation (2.110) again, it is found that

$$a_0 = 0 . \tag{2.113}$$

Equation (2.109) can be rewritten as

$$\delta_{i_1 i_4}(a_{i_1} + \delta_{i_2 i_3}) + \delta_{i_1 i_4}(a_{i_1} + \delta_{i_2 i_3}) + \delta_{i_1 i_4}(a_{i_1} + \delta_{i_2 i_3}) = 0, \tag{2.114}$$

from which the coefficients a_{i_1}, a_{i_2}, and a_{i_3} are found to be,

$$
\begin{aligned}
a_{i_1} &= -\delta_{i_2 i_3} \\
a_{i_2} &= -\delta_{i_1 i_3} \\
a_{i_3} &= -\delta_{i_1 i_2} .
\end{aligned}
\tag{2.115}
$$

The third order Polynomial Chaos can then be written as

$$\Gamma_3(\xi_{i_1}, \xi_{i_2}, \xi_{i_3}) = \xi_{i_1} \xi_{i_2} \xi_{i_3} - \xi_{i_1} \delta_{i_2 i_3} - \xi_{i_2} \delta_{i_1 i_3} - \xi_{i_3} \delta_{i_1 i_2} . \tag{2.116}$$

After laborious algebraic manipulations, the fourth order Polynomial Chaos can be expressed as

$$
\begin{aligned}
\Gamma_4(\xi_{i_1}, \xi_{i_2}, \xi_{i_3}, \xi_{i_4}) = \ & \xi_{i_1} \xi_{i_2} \xi_{i_3} \xi_{i_4} \\
- \ & \xi_{i_1} \xi_{i_2} \delta_{i_3 i_4} - \xi_{i_1} \xi_{i_3} \delta_{i_2 i_4} - \xi_{i_1} \xi_{i_4} \delta_{i_2 i_3} \\
- \ & \xi_{i_2} \xi_{i_3} \delta_{i_1 i_4} - \xi_{i_2} \xi_{i_4} \delta_{i_1 i_3} - \xi_{i_3} \xi_{i_4} \delta_{i_1 i_2} \\
+ \ & \delta_{i_1 i_2} \delta_{i_3 i_4} + \delta_{i_1 i_3} \delta_{i_2 i_4} + \delta_{i_1 i_4} \delta_{i_2 i_3} .
\end{aligned}
\tag{2.117}
$$

It is readily seen that, in general, the nth order Polynomial Chaos can be written as

$$\Gamma_p(\xi_{i_1}, ..., \xi_{i_n}) = \begin{cases} \sum_{\substack{r=n \\ r \ even}}^{0} (-1)^r \sum_{\pi(i_1,...,i_n)} \prod_{k=1}^{r} \xi_{i_k} < \prod_{l=r+1}^{n} \xi_{i_l} > \\ \\ \qquad\qquad n \ even \\ \\ \sum_{\substack{r=n \\ r \ even}}^{0} (-1)^{r-1} \sum_{\pi(i_1,...,i_n)} \prod_{k=1}^{r} \xi_{i_k} < \prod_{l=r+1}^{n} \xi_{i_l} > \\ \\ \qquad\qquad n \ odd \end{cases} \qquad (2.118)$$

where $\pi(.)$ denotes a permutation of its arguments, and the summation is over all such permutations such that the sets $\{\xi_{i_1}, ..., \xi_{i_r}\}$ is modified by the permutation.

Note that the Polynomial Chaoses as obtained in equations (2.89), (2.90), (2.98), (2.117) and (2.118) are orthogonal with respect to the Gaussian probability measure, which makes them identical with the corresponding multidimensional Hermite polynomials (Grad, 1949). These polynomials have been used extensively in relation to problems in turbulence theory (Imamura et.al, 1965a-b). This equivalence is implied by the orthogonality of the Polynomial Chaoses with respect to the inner product defined by equation (1.8) where dP is the Gaussian measure $e^{-\frac{1}{2}\boldsymbol{\xi}^T \boldsymbol{\xi}} d\boldsymbol{\xi}$ where $\boldsymbol{\xi}$ denotes the vector of n random variables $\{\xi_{i_k}\}_{k=1}^{n}$. This measure is exactly the weighing function with respect to which the Hermite polynomials are orthogonal in the L_2 sense (Oden, 1979). This fact suggests another method for constructing the Polynomial Chaoses, namely from the generating function of the Hermite polynomials. Thus, the Polynomial Chaos of order n can be obtained as

$$\Gamma_n(\xi_{i_1}, ..., \xi_{i_n}) = (-1)^n \frac{\partial^n}{\partial \xi_{i_1} ... \partial \xi_{i_n}} e^{-\frac{1}{2}\boldsymbol{\xi}^T \boldsymbol{\xi}} \qquad (2.119)$$

Equation (2.119) can be readily evaluated symbolically using the symbolic manipulation program MACSYMA (1986). Tables (2.1) - (2.4) display expressions for the one, two, three and four-dimensional Polynomial Chaoses up to the fourth order along with the values of their variances. The term Ψ_i in these tables refers to the quantity appearing in equation (2.86).

j	p, Order of the Homogeneous Chaos	j^{th} Polynomial Chaos Ψ_j	$<\Psi_j^2>$
0	$p = 0$	1	1
1	$p = 1$	ξ_1	1
2	$p = 2$	$\xi_1^2 - 1$	2
3	$p = 3$	$\xi_1^3 - 3\xi_1$	6
4	$p = 4$	$\xi_1^4 - 6\xi_1^2 + 3$	24

Table 2.1: One-Dimensional Polynomial Chaoses and Their Variances; $n = 1$

j	p, Order of the Homogeneous Chaos	j^{th} Polynomial Chaos Ψ_j	$<\Psi_j^2>$
0	$p = 0$	1	1
1	$p = 1$	ξ_1	1
2		ξ_2	1
3	$p = 2$	$\xi_1^2 - 1$	2
4		$\xi_1\xi_2$	1
5		$\xi_2^2 - 1$	2
6	$p = 3$	$\xi_1^3 - 3\xi_1$	6
7		$\xi_1^2\xi_2 - \xi_2$	2
8		$\xi_1\xi_2^2 - \xi_1$	2
9		$\xi_2^3 - 3\xi_2$	6
10	$p = 4$	$\xi_1^4 - 6\xi_1^2 + 3$	24
11		$\xi_1^3\xi_2 - 3\xi_1\xi_2$	6
12		$\xi_1^2\xi_2^2 - \xi_1^2 - \xi_2^2 + 1$	4
13		$\xi_1\xi_2^3 - 3\xi_1\xi_2$	6
14		$\xi_2^4 - 6\xi_2^2 + 3$	24

Table 2.2: Two-Dimensional Polynomial Chaoses and Their Variances; $n = 2$

j	p, Order of the Homogeneous Chaos	j^{th} Polynomial Chaos Ψ_j	$<\Psi_j^2>$
0	$p = 0$	1	1
1	$p = 1$	ξ_1	1
2		ξ_2	1
3		ξ_3	1
4	$p = 2$	$\xi_1^2 - 1$	2
5		$\xi_1\xi_2$	1
6		$\xi_1\xi_3$	1
7		$\xi_2^2 - 1$	2
8		$\xi_2\xi_3$	1
9		$\xi_3^2 - 1$	2
10	$p = 3$	$\xi_1^3 - 3\xi_1$	6
11		$\xi_1^2\xi_2 - \xi_2$	2
12		$\xi_1^2\xi_3 - \xi_3$	2
13		$\xi_1\xi_2^2 - \xi_1$	2
14		$\xi_1\xi_2\xi_3$	1
15		$\xi_1\xi_3^2 - \xi_1$	2
16		$\xi_2^3 - 3\xi_2$	6
17		$\xi_2^2\xi_3 - \xi_3$	2
18		$\xi_2\xi_3^2 - \xi_2$	2
19		$\xi_3^3 - 3\xi_3$	6
20	$p = 4$	$\xi_1^4 - 6\xi_1^2 + 3$	24
21		$\xi_1^3\xi_2 - 3\xi_1\xi_2$	6
22		$\xi_1^3\xi_3 - 3\xi_1\xi_3$	6
23		$\xi_1^2\xi_2^2 - \xi_2^2 - \xi_1^2 + 1$	4
24		$\xi_1^2\xi_2\xi_3 - \xi_2\xi_3$	2
25		$\xi_1^2\xi_3^2 - \xi_3^2 - \xi_1^2 + 1$	4
26		$\xi_1\xi_2^3 - 3\xi_1\xi_2$	6
27		$\xi_1\xi_2^2\xi_3 - \xi_1\xi_3$	2
28		$\xi_1\xi_2\xi_3^2 - \xi_1\xi_2$	2
29		$\xi_1\xi_3^3 - 3\xi_1\xi_3$	6
30		$\xi_2^4 - 6\xi_2^2 + 3$	2
31		$\xi_2^3\xi_3 - 3\xi_2\xi_3$	6
32		$\xi_2^2\xi_3^2 - \xi_3^2 - \xi_2^2 + 1$	4
33		$\xi_2\xi_2^3 - 3\xi_2\xi_3$	6
34		$\xi_3^4 - 6\xi_3^2 + 3$	24

Table 2.3: Three-Dimensional Polynomial Chaoses and Their Variances; $n = 3$

j	p, Order of the Homogeneous Chaos	j^{th} Polynomial Chaos Ψ_j	$<\Psi_j^2>$
0	$p = 0$	1	1
1	$p = 1$	ξ_1	1
2		ξ_2	1
3		ξ_3	1
4		ξ_4	1
5	$p = 2$	$\xi_1^2 - 1$	2
6		$\xi_1\xi_2$	1
7		$\xi_1\xi_3$	1
8		$\xi_1\xi_4$	1
9		$\xi_2^2 - 1$	2
10		$\xi_2\xi_3$	1
11		$\xi_2\xi_4$	1
12		$\xi_3^2 - 1$	2
13		$\xi_3\xi_4$	1
14		$\xi_4^2 - 1$	2
15	$p = 3$	$\xi_1^3 - 3\xi_1$	6
16		$\xi_1^2\xi_2 - \xi_2$	2
17		$\xi_1^2\xi_3 - \xi_3$	2
18		$\xi_1^2\xi_4 - \xi_4$	2
19		$\xi_1\xi_2^2 - \xi_1$	2
20		$\xi_1\xi_2\xi_3$	1
21		$\xi_1\xi_2\xi_4$	1
22		$\xi_1\xi_3^2 - \xi_1$	2
23		$\xi_1\xi_3\xi_4$	1
24		$\xi_1\xi_4^2 - \xi_4$	2
25		$\xi_2^3 - 3\xi_2$	6
26		$\xi_2^2\xi_3 - \xi_3$	2
27		$\xi_2^2\xi_4 - \xi_4$	2
28		$\xi_2\xi_3^2 - \xi_2$	2
29		$\xi_2\xi_3\xi_4$	1
30		$\xi_2\xi_4^2 - \xi_2$	2
31		$\xi_3^3 - 3\xi_3$	6
32		$\xi_3^2\xi_4 - \xi_4$	2
33		$\xi_3\xi_4^2 - \xi_3$	2
34		$\xi_4^3 - 3\xi_4$	6

Table 2.4: Four-Dimensional Polynomial Chaoses and Their Variances; $n = 4$

i	p, Order of the Homogeneous Chaos	i^{th} Polynomial Chaos Ψ_i	$<\Psi_i^2>$
35	$p = 4$	$\xi_1^4 - 6\xi_1^2 + 3$	24
36		$\xi_1^3\xi_2 - 3\xi_1\xi_2$	6
37		$\xi_1^3\xi_3 - 3\xi_1\xi_3$	6
38		$\xi_1^3\xi_4 - 3\xi_1\xi_4$	6
39		$\xi_1^2\xi_2^2 - \xi_1^1 - \xi_2^2 + 1$	4
40		$\xi_1^2\xi_2\xi_3 - \xi_2\xi_3$	2
41		$\xi_1^2\xi_2\xi_4 - \xi_2\xi_4$	2
42		$\xi_1^2\xi_3^2 - \xi_3^2 - \xi_1^2 + 1$	4
43		$\xi_1^2\xi_3\xi_4 - \xi_3\xi_4$	2
44		$\xi_1^2\xi_4^2 - \xi_4^2 - \xi_1^2 + 1$	4
45		$\xi_1\xi_2^3 - 3\xi_1\xi_2$	6
46		$\xi_1\xi_2^2\xi_3 - \xi_1\xi_3$	2
47		$\xi_1\xi_2^2\xi_4 - \xi_1\xi_4$	2
48		$\xi_1\xi_2\xi_3^2 - \xi_1\xi_2$	2
49		$\xi_1\xi_2\xi_3\xi_4$	1
50		$\xi_1\xi_2\xi_4^2 - \xi_1\xi_2$	2
51		$\xi_1\xi_3^3 - 3\xi_1\xi_3$	6
52		$\xi_1\xi_3^2\xi_4 - \xi_1\xi_4$	2
53		$\xi_1\xi_3\xi_4^2 - \xi_1\xi_3$	2
54		$\xi_1\xi_4^3 - 3\xi_1\xi_4$	6
55		$\xi_2^4 - 6\xi_2^2 + 3$	24
56		$\xi_2^3\xi_3 - 3\xi_2\xi_3$	6
57		$\xi_2^3\xi_4 - 3\xi_2\xi_4$	6
58		$\xi_2^2\xi_3^2 - \xi_3^2 - \xi_2^2 + 1$	
59		$\xi_2^2\xi_3\xi_4 - \xi_3\xi_4$	2
60		$\xi_2^2\xi_4^2 - \xi_4^2 - \xi_2^2 + 1$	4
61		$\xi_2\xi_3^3 - 3\xi_2\xi_3$	6
62		$\xi_2\xi_3^2\xi_4 - \xi_2\xi_4$	2
63		$\xi_2\xi_3\xi_4^2 - \xi_2\xi_3$	2
64		$\xi_2\xi_4^3 - 3\xi_2\xi_4$	6
65		$\xi_3^4 - 6\xi_3^2 + 3$	24
66		$\xi_3^3\xi_4 - 3\xi_3\xi_4$	6
67		$\xi_3^2\xi_4^2 - \xi_3^2 - \xi_4^2 + 1$	4
68		$\xi_3\xi_4^3 - 3\xi_3\xi_4$	6
69		$\xi_4^4 - 6\xi_4^2 + 3$	24

Table 2.4 (continued): Four-Dimensional Polynomial Chaoses and Their Variances; $n = 4$

The expressions corresponding to these polynomials were obtained using the MACSYMA programs shown in Figure (2.16) for the two-dimensional polynomials and Figure (2.17) for the four-dimensional polynomials. The first two terms in equation (2.85) represent the Gaussian component of the function $\mu(\theta)$. Therefore, for a Gaussian process, this expansion reduces to a single summation, the coefficients a_{i_1} being the coefficients in the Karhunen-Loeve expansion of the process. Note that equation (2.85) is a convergent series representation for the functional operator $h[.]$ appearing in equation (2.83). For a given non-Gaussian process defined by its probability distribution function, a representation in the form given by equation (2.85) can be obtained by projecting the process on the successive Homogeneous Chaoses. This can be achieved by using the inner product defined by equation (1.8) to determine the requisite coefficients. This concept has been successfully applied in devising efficient variance reduction techniques to be coupled with the Monte Carlo simulation method (Chorin, 1971; Maltz and Hitzl, 1979).

```
* kronecker : *\

kdelta(x,y)::=buildq([x,y],
if x=y then 1 else 0)$

\* average : *\

average(p)::=
buildq([p],
( logexpand:super,
norder:0,
for i: 1 thru n
do  ( z:coeff(log(p),log(x[i])),
if z#0 then
for j:1 thru z do y[j+order]:x[i],
order:order+z ),
hom:p/prod(y[i],i,1,order),
if oddp(order) then 0 else
(if order=0 then
hom
else
if order=2 then
hom*kdelta(y[1],y[2])
else
hom*sum(kdelta(y[1],y[i])*prod(y[k],k,2,order)
        /y[i],i,2,order))))$
```

Table 2.5:, MACSYMA Macro to Generate the Two-Dimensional
Polynomial Chaoses and Evaluate Their Variance.

```
\* herm2 : *\

ind:0$
G2[0]:1$
gen:exp(-sum(x[i]^2/2,i,1,2))$
for i:1 thru 2 do
(ind:ind+1,
G2[ind]:expand(-diff(gen,x[i])/gen) )$
for i:1 thru 2 do
(for j:i thru 2 do
(ind:ind+1,
G2[ind]:expand(-diff(-diff(gen,x[i]),x[j])/gen)))$
for i:1 thru 2 do
(for j:i thru 2 do
(for k:j thru 2 do
(ind:ind+1,
G2[ind]:expand(-diff(-diff(
        -diff(gen,x[i]),x[j]),x[k])/gen))))$
for i:1 thru 2 do
(for j:i thru 2 do
(for k:j thru 2 do
(for l:k thru 2 do
(ind:ind+1,
G2[ind]:expand(-diff(-diff(-diff(
        -diff(gen,x[i]),x[j]),x[k]),x[l])/gen )))))$
```

Table 2.5, (continued):, MACSYMA Macro to Generate the Two-Dimensional
Polynomial Chaoses and Evaluate Their Variance.

```
\* square2 : *\

square2(num)::=(buildq([num],
(
n:2,
load(kronecker),
load(average),
load(herm2),
for j:1 thru num do
(pp:expand(G2[j]*G2[j]),
p[0]:pp,
mm:0,
for m:1 while integerp(p[m-1])=false do
(mm:m,
p[m]:if length(p[m-1])=2
and integerp(part(p[m-1],2))=true
and length(part(p[m-1],1))=1
then average(p[m-1])
else map(average,p[m-1])),
var2[j]:p[mm],
print(j,var2[j],pp) ) )))$
```

Table 2.5, (continued):, MACSYMA Macro to Generate the Two-Dimensional
Polynomial Chaoses and Evaluate Their Variance.

```
\* kronecker : *\

kdelta(x,y)::=buildq([x,y],
if x=y then 1 else 0)$

\* average : *\

average(p)::=
buildq([p],
( logexpand:super,
order:0,
for i: 1 thru n
do ( z:coeff(log(p),log(x[i])),
if z#0 then
for j:1 thru z do y[j+order]:x[i],
order:order+z ),
hom:p/prod(y[i],i,1,order),
if oddp(order) then 0 else
(if order=0 then
hom
else
if order=2 then
hom*kdelta(y[1],y[2])
else
hom*sum(kdelta(y[1],y[i])*prod(y[k],k,2,order)
        /y[i],i,2,order))))$
```

Table 2.6:, MACSYMA Macro to Generate the Four-Dimensional
Polynomial Chaoses and Evaluate Their Variance.

```
\* herm4 : *\

ind:0$
G4[0]:1$
gen:exp(-sum(x[i]^2/2,i,1,4))$
for i:1 thru 4 do
(ind:ind+1,
G4[ind]:expand(-diff(gen,x[i])/gen) )$
for i:1 thru 4 do
(for j:i thru 4 do
(ind:ind+1,
G4[ind]:expand(-diff(-diff(gen,x[i]),x[j])/gen) ))$
for i:1 thru 4 do
(for j:i thru 4 do
(for k:j thru 4 do
(ind:ind+1,
G4[ind]:expand(-diff(-diff(
        -diff(gen,x[i]),x[j]),x[k])/gen))))$
for i:1 thru 4 do
(for j:i thru 4 do
(for k:j thru 4 do
(for l:k thru 4 do
(ind:ind+1,
G4[ind]:expand(-diff(-diff(-diff(
        -diff(gen,x[i]),x[j]),x[k]),x[l])/gen) ))))$
```

Table 2.6 (continued):, MACSYMA Macro to Generate the Four-Dimensional
Polynomial Chaoses and Evaluate Their Variance.

```
\* square4 : *\

square4(num)::=(buildq([num],
(
n:4,
load(kronecker),
load(average),
load(herm4),
for j:1 thru num do
(pp:expand(G4[j]*G4[j]),
p[0]:pp,
for m:1 while integerp(p[m-1])=false do
(mm:m,
p[m]:if length(p[m-1])=2
and integerp(part(p[m-1],2))=true
and length(part(p[m-1],1))=1
then average(p[m-1])
else map(average,p[m-1])),
var4[j]:p[mm],
print(j,var4[j],pp) ) )))$
```

Table 2.6 (continued):, MACSYMA Macro to Generate the Four-Dimensional
Polynomial Chaoses and Evaluate Their Variance.

Chapter 3

STOCHASTIC FINITE ELEMENT METHOD: Response Representation

3.1 Preliminary Remarks

Recent analytical methods for addressing the problem of system stochasticity in the context of structural mechanics have involved either solving for the second order statistics of the response or implementing a First-Order or a Second-Order (FORM/SORM) (Der-Kiureghian et.al, 1987a) algorithm in conjunction with a finite element code. Monte Carlo simulation methods have also been investigated towards improving their efficiency and versatility (Shinozuka and Astill, 1972; Shinozuka and Lenoe, 1976; Polhemus and Cakmak, 1981; Cakmak and Sherif, 1984; Shinozuka, 1987; Spanos and Mignolet, 1987). However, due to the substantial requisite computational effort, the Monte Carlo simulation method is still used mainly to verify other approaches. The perturbation method (Nakagiri and Hisada, 1982; Liu, Besterfield and Belytschko, 1988) and the Neumann expansion method (Shinozuka and Nomoto, 1980; Adomian and Malakian, 1980; Benaroya and Rehak, 1987; Shinozuka, 1987) have been developed and shown to provide acceptable results for small random fluctuations in the material properties. Typical results from these two methods have been restricted to second order statistics of the response quantities. Obviously, this information does not suffice for a meaningful risk assessment and failure analysis. In a different context, a stochastic finite element method has been developed that couples response surface techniques with deterministic finite element formulations. The emphasis in these methods has been on developing efficient search algorithms for locating the point, termed the design point, at which the response surface is to be expanded in a first or second order Taylor series (Der-Kiureghian et.al, 1986, 1987, Cruse et.al, 1988) and on formulating the problem using equivalent Gaussian variables (Wu, 1987; Wirshing and Wu 1987; Wu and Wirshing, 1987). Following this approach, a number of finite element simulations are performed, the solutions of which define an approximation to the response surface. From this approximation, the design point is subsequently determined. Once located, the distance from this point to the origin can be used to obtain a rough approximation to the probability of failure. The accuracy of this approximation deteriorates rapidly as the dimension of the space increases. A number of stochastic finite element codes have emerged from these research efforts. They all utilize a deterministic finite element code which is coupled with either a standard optimization algorithm or a search algorithm that is customized to implement the specific reliability concept being adopted. Of interest are the codes NESSUS/EXPERT, being developed at the Southwest Research Institute in collaboration with NASA-Lewis Research Center (Millwater et.al, 1988; Millwater et.al 1989; Cruse and Chamis, 1989), and CALREL-FEAP, being developed at UC-Berkeley (Liu,et.al, 1989). Note that alternative methods to describe the response surface that do not rely on a Taylor series expansion have been proposed in the literature. Of these, the polynomial approximation suggested by Grigoriu (1982) is worth mentioning. It relies on fitting a polynomial to a number of points on the response surface. Each one of these points, however, is obtained as the solution to a finite element simulation.

Alternatively, methods based on the optimal feature extraction properties of the Karhunen-Loeve have been recently added to the arsenal of approaches to address the associated class of problems. The Karhunen-Loeve decomposition has been coupled with either a Neumann expansion scheme (Spanos and Ghanem, 1989) or a Polynomial Chaos expansion in conjunction with a Galerkin projection (Ghanem and Spanos, 1990) to achieve an efficient implementation of the randomness into the solution procedure.

In this chapter the deterministic finite element method is first briefly reviewed. Then, the non-spectral methods for solving a class of stochastic problems are discussed. Finally, two spectral methods involving the Karhunen-Loeve expansion and the concept of Polynomial Chaos for stochastic finite element analysis are examined in detail. It is noted that the prime theme of this chapter is an expeditious representation of the system response itself; the determination of the statistics of the response will be addressed in Chapter IV.

3.2 Deterministic Finite Elements

3.2.1 Problem Definition

In the deterministic case, the space Ω of elementary events is reduced to a single element, coinciding with the actual realization of the problem. It is obvious, then, that this is a special case of the problem defined in Chapter I. For convenience, and omitting the randomness argument θ, equation (1.12) is rewritten as

$$\mathbf{L}(\mathbf{x})\left[u(\mathbf{x})\right] = f(\mathbf{x}), \quad \mathbf{x} \in \mathbf{D}. \tag{3.1}$$

In the following, two equivalent formulations for deterministic finite element analysis are presented. They are the variational formulation and the Galerkin formulation. The stochastic finite element formulation as introduced in the sequel is based on a Galerkin projection in the space $\mathbf{\Theta}$ of random variables. However, the variational formulation is presented to draw a physically appealing analogy between the mechanical energy of a system as given by its potential energy, and the uncertainty energy of a system, as given by its information-entropy. Note that the exposition in this section is meant to outline some of the basic concepts of the deterministic finite element method that are relevant to the stochastic case; it is not meant to provide an account of the state-of-the-art of finite element techniques.

Equation (3.1) can be viewed as a mapping from the space over which the response $u(\mathbf{x})$ is defined to the space over which the excitation $f(\mathbf{x})$ is defined. All the excitations dealt with in the sequel, as well as in most engineering applications, have finite energy. That is,

$$\int_{\mathbf{D}} f(\mathbf{x})^2 \, dx < \infty. \tag{3.2}$$

The range of the operator $\mathbf{L}(\mathbf{x})$ is thus the space of all square-integrable functions. The domain of $\mathbf{L}(\mathbf{x})$, that is the space spanned by the solution, is obviously determined by the special form of $\mathbf{L}(\mathbf{x})$, as well as by the boundary conditions and the initial conditions associated with the physical problem. Assuming that $\mathbf{L}(\mathbf{x})$ is an mth order differential operator, let $\mathbf{C}^m$ denote the space of all functions that are m times differentiable. Then, the solution space is some subspace of $\mathbf{C}^m$ whose elements satisfy the associated homogeneous essential boundary conditions of the problem. The exact nature of this subspace depends on the specific finite element formulation employed and will not be dwelled upon any further. The solution function $u(\mathbf{x})$ can be expanded along a basis in this space, and equation (3.1) becomes

$$\mathbf{L}(\mathbf{x}) \left[\sum_{i=1}^{\infty} u_i \, g_i(\mathbf{x}) \right] = \sum_{i=1}^{\infty} u_i \, \mathbf{L}(\mathbf{x}) \, [g_i(\mathbf{x})]$$
$$= f(\mathbf{x}) , \qquad (3.3)$$

where u_i is the component of the solution $u(\mathbf{x})$ along the basis function $g_i(\mathbf{x})$. In the sequel, the above summations are truncated at the Nth term and the problem then is to compute the coordinates $\{u_i\}$ of the response $u(\mathbf{x})$ with respect to the finite-dimensional basis $\{g_i(\mathbf{x})\}_{i=1}^{N}$.

3.2.2   Variational Approach

Variational principles were introduced and studied well before the introduction of the finite element method. The theory has become an integral part of functional analysis with a solid mathematical foundation. The physical meaning of the corresponding minimization problem, although quite helpful, is no longer necessary, as long as the operator describing the system satisfies certain conditions of self-adjointness (Rektorys, 1980). In fact, it may be shown that to every self-adjoint operator equation, is associated a quadratic functional whose stationary value coincides with the solution of the equation. Physically speaking, a self-adjoint operator results from situations where reciprocity, as given by Betti's law (Shames and Dym, 1985) for example, is applicable. Such is the case with most linear differential operators of even order. Then, the solution to equation (3.1) is that function $u(\mathbf{x})$ which minimizes the functional

$$\mathbf{I}[v(\mathbf{x})] = (\, \mathbf{L}(\mathbf{x}) \, [v(\mathbf{x})] \, , \, v(\mathbf{x}) \,) - 2 \, (\, f(\mathbf{x}) \, , \, v(\mathbf{x}) \,) , \qquad (3.4)$$

where $(\; . \quad , \; . \;)$ denotes a suitable inner product. Indeed, setting the functional derivative of $\mathbf{I}[v(\mathbf{x})]$ with respect to $v(\mathbf{x})$ to zero, results in

$$\mathbf{L}(\mathbf{x}) [v(\mathbf{x})] - f(\mathbf{x}) = 0 . \tag{3.5}$$

Physically, the functional $\mathbf{I}[v(\mathbf{x})]$ corresponds to a measure of the energy in the system, which assumes its stationary value at the true response. Substituting equation (3.3) into equation (3.4) gives

$$\mathbf{I}[v(\mathbf{x})] = \sum_{i=1}^{N} \sum_{j=1}^{N} [v_i v_j (\mathbf{L}(\mathbf{x}) [g_i(\mathbf{x})] , g_j(\mathbf{x}))] \tag{3.6}$$
$$- 2 \sum_{i=1}^{N} v_i (f(\mathbf{x}) , g_i(\mathbf{x})) .$$

Setting the variation of $\mathbf{I}[v(\mathbf{x})]$ to zero, and recognizing that $u(\mathbf{x}) = v(\mathbf{x})$ at the stationary point of $\mathbf{I}[.]$ gives,

$$\frac{\partial \mathbf{I}[v(\mathbf{x})]}{\partial v_j}\Big|_{v(\mathbf{x})=u(\mathbf{x})} = 2 \sum_{i=1}^{N} (\mathbf{L}(\mathbf{x}) [g_i(\mathbf{x})] , g_j(\mathbf{x})) u_j$$
$$- 2 (f(\mathbf{x}) , g_j(\mathbf{x}))$$
$$= 0 , j = 1, ..., N . \tag{3.7}$$

Equation (3.7) can be expressed as a system of algebraic equations,

$$\mathbf{L} \, \mathbf{u} = \mathbf{f} \tag{3.8}$$

where

$$\mathbf{L}_{ij} = (\mathbf{L}(\mathbf{x}) [g_i(\mathbf{x})] , g_j(\mathbf{x})) \tag{3.9}$$

$$\mathbf{f}_i = (f(\mathbf{x}) , g_i(\mathbf{x})) . \tag{3.10}$$

Note that as given by equation (3.9), the basis $\{g_i(\mathbf{x})\}_{i=1}^{N}$ must span an N-dimensional subspace of $\mathbf{C}^m$. At this point, it may be noted that performing the integration indicated by equation (3.9) by parts, the operator $\mathbf{L}(\mathbf{x})[.]$ can be split into two lower order operators as in

$$\mathbf{L}_{ij} = (\mathbf{L}_1(\mathbf{x}) [g_i(\mathbf{x})] , \mathbf{L}_2(\mathbf{x}) [g_j(\mathbf{x})]) , \tag{3.11}$$

where now $\mathbf{L}_1[.]$ and $\mathbf{L}_2[.]$ are two appropriate operators of lower order than $\mathbf{L}(\mathbf{x})$.

Whether equation (3.9) or equation (3.11) is used to compute $\mathbf{L}_{ij}$, the result is an equation of the form (3.8). Ordinarily, the set $g_i(\mathbf{x})$ is chosen to be the set of piecewise polynomials of order $n + 1$. This choice for the basis set generates a natural discretization of the domain $\mathbf{D}$.

The variational approach just outlined is used later in this chapter to generate a set of algebraic equations from the governing differential equation. The main point to carry over from this section to the stochastic case is that the solution of the problem can be obtained as the element of a certain Hilbert space of admissible functions that minimizes a certain norm of the energy of the system. In Chapter IV, an analogy is drawn to this principle for the stochastic case, where the energy norm is replaced by an uncertainty norm.

3.2.3   Galerkin Approach

A condition associated with the variational formulation of the finite element method is that the operator $\mathbf{L}(\mathbf{x})[.]$ be self-adjoint. This restriction excludes a class of important practical problems. Further, in some cases, the variational principle for a given problem may lack the intuitive physical association with the energy principle. Thus, it may be advantageous or even necessary to resort to the Galerkin formulation of the finite element method (Zienkiewicz and Taylor, 1989). The method consists of expanding the response function along a basis of a finite dimensional subspace of an admissible Hilbert space (Oden, 1979), and requiring that the error resulting from taking a finite number of terms in the expansion be orthogonal to another Hilbert space, the "test space", whose functions are called "test functions". Usually, the test space is chosen to coincide with the admissible space. In other words, expanding $u(\mathbf{x})$ in equation (3.1) in terms of a finite dimensional basis $\{g_i(\mathbf{x})\}_{i=1}^{N}$ of a subspace of the space $\mathbf{C}^m$ introduces an error of the form

$$\epsilon_L = \sum_{i=1}^{N} \mathbf{L}(\mathbf{x}) \, [g_i(\mathbf{x})] \, - \, f(\mathbf{x}) \, . \tag{3.12}$$

Requiring this error to be orthogonal to the subspace spanned by $\{g_i(\mathbf{x})\}_{i=1}^N$, yields a set of N algebraic equations

$$(\epsilon_L , g_j(\mathbf{x})) = 0 \; j = 1, ..., N . \tag{3.13}$$

This equation is equivalent to

$$\sum_{i=1}^N u_i (\mathbf{L}(\mathbf{x})[g_i(\mathbf{x})], g_j(\mathbf{x})) = (f(\mathbf{x}), g_j(\mathbf{x})), \; j = 1, ..., N, \tag{3.14}$$

which is similar to equation (3.8). Here again, integration by parts can be used to enlarge the admissible space. The Galerkin method will form the basis for the spectral extension of the finite element method to problems involving stochastic operators.

3.2.4 p-Adaptive Methods, Spectral Methods and Hierarchical Finite Element Bases

It is customary in finite element analysis to regard the coefficients of the basis functions as representing physically meaningful nodal variables. New degrees of freedom are introduced through mesh refinement with the new coefficients representing the physical quantities at the new nodal points. Obviously, as more degrees of freedom are introduced, the approximation error diminishes. This is the so-called h-method whereby the approximation error is reduced through successive mesh refinement. This method constitutes the most widely used version of the finite element method. Recently, however, a new error reduction techniques with better convergence properties has been actively investigated. The so-called p-method consists of reducing the approximation error by using higher order interpolation within each element (Babuska, I. et.al, 1986). Thus, the same discrete mesh is used in successive approximations. The successive interpolation bases over each element are introduced in a hierarchical manner permitting results from previous approximations to be efficiently used in computing higher order approximations. In the limit, as the number of finite elements is reduced to one, a global approximation is obtained. This limiting case is often referred to as a spectral approximation. It has some obvious advantages when dealing with functions that are defined over domains that are either too difficult to discretize or too abstract for such a discretization to be intuitively feasible. Indeed, the space of random variables is just such a space. The measure associated with this space is a probability measure that cannot be made to correspond with the Lebesgue measure used in spatial discretization procedures. Still, functions defined over this space can be discretized using an adequate spectral measure. In this context, discretization refers to expanding the functions using a set of basis functions that are globally defined. An additional appeal of the spectral formulation is the fact that global approximations can make efficient use of hierarchical basis functions. Interestingly, the set of basis functions used in the ensuing spectral stochastic finite element formulation is hierarchical and the associated solution procedure can be optimized to make use of this fact.

3.3   Stochastic Finite Elements

3.3.1   Preliminary Remarks

In this section three of the methods used for treating problems involving random media are first discussed. Then, two recently developed methods are presented. The three methods have been selected due to their popularity with various investigators and their compatibility with the finite element method. The theoretical foundation and the numerical implementation for each of these methods are discussed.

Examining realizations of a random process, two important features can be observed. The frequency content of the random fluctuations and the magnitude of the fluctuations. The former feature can be used to situate the random process with respect to white noise. The broader the frequency content, the closer is the process to white noise (Lin, 1967). This feature reflects the level of correlation of the process at two points in its domain. The second feature reflects to, a certain extent, the degree of uncertainty associated with the process and can be related to its coefficient of variation. In view of that, a meaningful definition of the range of applicability of a given method is the range of coefficients of variations and the range of frequency content that the method can accommodate. All three methods to be discussed are based on performing direct operations on equation (1.12). To show explicitly the dependence of $\Pi[.]$ on the random process $\alpha(\mathbf{x}, \theta)$ this equation can be rewritten as

$$[\mathbf{L}(\mathbf{x}) + \Pi(\alpha(\mathbf{x}, \theta), \mathbf{x})] [u(\alpha(\mathbf{x}, \theta), \mathbf{x})] = f(\mathbf{x}, \theta), \qquad (3.15)$$

3.3.2 Monte Carlo Simulation (MCS)

The usefulness of the Monte Carlo Simulation method (MCS) is based on the fact that the next best situation to having the probability distribution function of a certain random quantity is to have a corresponding large population. The implementation of the method consists of numerically simulating a population corresponding to the random quantities in the physical problem, solving the deterministic problem associated with each member of that population, and obtaining a population corresponding to the random response quantities. This population can then be used to obtain statistics of the response variables.

The Monte Carlo method is a quite versatile mathematical tool capable of handling situations where all other methods fail to succeed. The method has been known and used extensively in various fields such as health care, agriculture, and econometrics. However, in engineering mechanics it has attracted intense attention only recently following the widespread availability of inexpensive computational systems. This computational availability has triggered, indeed, an interest in developing sophisticated and efficient simulation algorithms. Shinozuka and Jan (1972) have had a pioneering role in introducing the method to the field of engineering mechanics. They have suggested simulating a random process as the superposition of a large number of sinusoids having a uniformly distributed random phase angle. This approach has been successfully used in a variety of problems for the simulation of earthquake records, sea-wave elevations, and various other random phenomena. Later, Shinozuka used an FFT algorithm in conjunction with the MCS to achieve a more efficient implementation of the simulation procedure (Shinozuka, 1974). Subsequent developments have involved the application of Wiener's filtering theory to problems of structural mechanics. The main idea here is to view the process to be realized as the output of a linear filter excited by white noise. The problem is then reduced to computing a set of coefficients that specify the filter which can be excited with simulated white noise to produce the desired realizations. Various optimality criteria can be used in the design of these filters (Rabiner and Gold, 1975; Spanos and Hansen, 1981; Spanos and Mignolet, 1986). Most of the uses of the MCS have been in the study of random vibrations of deterministic media. For a review, see Spanos and Mignolet (1989). One of the first applications involving a random medium was presented by Shinozuka and Lenoe (1976) whereby a two-dimensional FFT algorithm was used to translate a two-dimensional random plate problem into a format compatible with the finite element method. Subsequent applications to problems of statics and dynamics involving random media were again carried out by Shinozuka and a number of co-workers (1980, 1985, 1986, 1987).

Over a period of years the application of the MCS to problems of structural mechanics has involved one-dimensional Gaussian processes. Recently, the digital simulation of non-Gaussian processes has been investigated (Yamazaki, 1987). Further, results have also been obtained for the simulation of two-dimensional processes (Mignolet, 1987). As far as using the MCS in problems involving random media is concerned, the computational cost of the approach is apparent. Ordinarily, the MCS is used as a brute force technique for assessing the validity of other approaches. It involves numerically generating realizations of the random processes $\alpha_k(\mathbf{x}, \theta)$ and $f(\mathbf{x}, \theta)$ appearing in equation (3.15), and proceeding with a deterministic equation to be solved for $u(\alpha(\mathbf{x}, \theta), \mathbf{x})$, for a fixed value of θ. The procedure is repeated a number of times for different values of $\theta \in \Omega$. This is obviously a collocation scheme in the space Ω of elementary random events. Realizations of the process are obtained, usually, either as local averages as discussed in section (2.2) or by representing the process as the response of a linear filter to a white noise excitation. The first approach is usually used when spatial randomness is involved, whereas the second approach is preferred for temporal random fluctuations.

3.3.3 Perturbation Method

The perturbation approach as applied to problems of random media is an extension of the method used in nonlinear analysis (Nayfeh, 1973; Jordan and Smith, 1977). Given certain smoothness conditions, the functions and operators involved can be expanded in a Taylor series about their respective mean values. To outline the method, it is assumed that the random process representing the system variability has been modeled by r random variables and that the excitation is deterministic. In this case, equation (3.15) becomes

$$[\ \mathbf{L}(\mathbf{x}) \ + \ \mathbf{\Pi}(\alpha(\theta), \mathbf{x}) \ ]\ [u(\alpha(\theta), \mathbf{x})] \ = \ f(\mathbf{x}) \qquad (3.16)$$

Expanding $\mathbf{\Pi}(\alpha(\theta), \mathbf{x})$ and $u(\alpha(\theta), \mathbf{x})$ about their mean values and noting that $\alpha_k(\theta)$ is a zero-mean random vector, leads to

$$\Pi(\alpha(\theta), \mathbf{x}) = \sum_{i=1}^{r} \alpha_i(\theta) \frac{\partial}{\partial \alpha_i(\theta)} \Pi(\alpha(\theta), \mathbf{x}) \tag{3.17}$$

$$+ \sum_{i=1}^{r} \sum_{j=1}^{r} \alpha_i(\theta) \, \alpha_j(\theta) \frac{\partial^2}{\partial \alpha_i(\theta) \partial \alpha_j(\theta)} \Pi(\alpha(\theta), \mathbf{x}) + \dots$$

and

$$u(\alpha(\theta), \mathbf{x}) = \bar{u}(\mathbf{x}) + \sum_{i=1}^{r} \alpha_i(\theta) \frac{\partial}{\partial \alpha_i(\theta)} u(\alpha(\theta), \mathbf{x}) \tag{3.18}$$

$$+ \sum_{i=1}^{r} \sum_{j=1}^{r} \alpha_i(\theta) \, \alpha_j(\theta) \frac{\partial^2}{\partial \alpha_i(\theta) \partial \alpha_j(\theta)} u(\alpha(\theta), \mathbf{x}) + \dots$$

Assuming small random deviations of the variables $\alpha_k(\theta)$ from their mean values, the contributions from all $\alpha_k(\theta)$ are small compared to that of the average quantity. Then, substituting back into equation (3.16), a multidimensional polynomial in $\alpha_k(\theta)$ is obtained. The right-hand side is obviously a polynomial of order zero in α. Equating same order polynomials on both sides yields a set of equations to be solved sequentially for the successive derivatives of the response. The first two terms of this succession are

$$\mathbf{L}(\mathbf{x}) \, [\bar{u}(\mathbf{x})] = f(\mathbf{x}) \tag{3.19}$$

$$\mathbf{L}(\mathbf{x}) \, [\frac{\partial}{\partial \alpha_i(\theta)} u(\alpha(\theta), \mathbf{x})] + \frac{\partial}{\partial \alpha_i(\theta)} \Pi(\alpha(\theta), \mathbf{x}) \, [\bar{u}(\mathbf{x})] = 0 \tag{3.20}$$

The larger the magnitude of the random fluctuations is, the more terms should be included in equations (3.17) and (3.18). This task is considerably complicated, thus greatly limiting the applicability of the method. Further, secular terms appear in higher order expansions. These are terms the magnitude of which increases with increasing approximation order, thus causing the expansion to diverge. The method has been applied extensively in recent years to problems involving random media. Nakagiri and Hisada (1982) and Hisada and Nakagiri (1985), and Liu et.al (1985, 1986, 1988) have applied the method to linear and nonlinear problems of statics and dynamics. Good results have been obtained for narrow band random fluctuations of small magnitude.

Once several of the sequential equations (3.19) and (3.20) have been solved, the response process can be written formally as

$$
\begin{aligned}
u(\alpha(\theta), \mathbf{x}) \;=\;& \bar{u}(\mathbf{x}) \;+\; \sum_{i=1}^{r} \alpha_i(\theta) \, \frac{\partial}{\partial \alpha_i(\theta)} u(\alpha(\theta), \mathbf{x}) \qquad\qquad (3.21) \\
&+\; \sum_{i=1}^{r} \sum_{j=1}^{r} \alpha_i(\theta) \, \alpha_j(\theta) \, \frac{\partial^2}{\partial \alpha_i(\theta) \, \partial \alpha_j(\theta)} u(\alpha(\theta), \mathbf{x}) \;+\; ...
\end{aligned}
$$

where all the partial derivatives are evaluated at the mean value (zero) of the random variables $\{\alpha_i(\theta)\}_{i=1}^{r}$. The case where the random aspect of the problem is modeled as a random process can be reduced to the previous case by recalling the discussion in Chapter II concerning local average representation of random processes. Namely, a random process can be replaced by the random variables corresponding to its spatial average over a number of subdomains in its domain of definition. Note that the number of random variables involved in this case is equal to the number of subdomains over which the averaging is performed. Usually these subdomains are the same as the elements used in a finite element analysis. The amount of computations required to perform the indicated operations becomes prohibitively large rather quickly as more terms are included in equations (3.17) and (3.18). For implementing the perturbation method into a finite element analysis, either the variational principle or the Galerkin approach can be used directly on equations (3.19) and (3.20), yielding a sequential system of algebraic equations to be solved for the successive derivatives of the response. Alternatively, these methods can be applied to equation (3.16), prior to performing the perturbation expansion, which when carried out, yields the same sequential algebraic equations as before. In terms of the comparison criterion defined at the beginning of this chapter, note that the higher the frequency of random fluctuations, the more random variables are needed to represent the process. That is, more elements are required and the problem can become quite large. It is noted at this point that the convergence of the series (3.21) is not guaranteed even for small levels of random fluctuations due to the presence of secular terms in higher order expansions. The first term in equation (3.21) is obviously the average response. The second moment of the response process can be obtained, at least conceptually, upon multiplying equation (3.21) with its conjugate, by averaging. The difficulties associated with that computation are such that only first order second moment statistics are usually computed; order here refers to the order of the polynomial in $\alpha_i(\mathbf{x}, \theta)$ included in the expansion. It is obvious that the method cannot be readily extended to compute the probability distribution function of the response process.

3.3.4 Neumann Expansion Method

From an operator-theoretic perspective, equation (3.15) reflects the problem of computing the inverse of a given operator. It is well known that when it exists, the inverse of an operator, called the resolvent, can be expanded in a convergent series in terms of the iterated kernels (Mikhlin, 1957). The theory was developed by Neumann and was further investigated by Fredholm (1903). Subsequent extensions of the theory include the concept of generalized inverse (Nashed, 1976). The concept was applied to the solution of stochastic operator equations by Bharrucha-Reid (1959). Subsequent contributions were made by Adomian (1983) and Adomian and Malakian (1980). The treatment was largely theoretical until Shinozuka and Nomoto (1980) introduced this concept to the field of structural mechanics. The method proved the implementation of higher order terms in the expansion to be quite laborious. Later, Shinozuka and co-workers coupled the Neumann expansion with the Monte Carlo simulation method to produce an efficient algorithm. The Neumann expansion method consists of expressing the solution to equation (3.15) as the series

$$u(\alpha(\theta), \mathbf{x}) = \sum_{i=0}^{\infty} (-1)^i \left[\mathbf{L}^{-1}(\mathbf{x}) \, \mathbf{\Pi}(\alpha(\mathbf{x}, \theta), \mathbf{x}) \right]^i [f(\mathbf{x}, \theta)] . \qquad (3.22)$$

To guarantee the convergence of series (3.22), it is necessary that the following criterion be satisfied,

$$\| \mathbf{L}^{-1}(\mathbf{x}) \, \mathbf{\Pi}(\alpha(\mathbf{x}, \theta), \mathbf{x}) \| < 1 . \qquad (3.23)$$

It is apparent that the algebraic manipulations incurred by including higher order terms, even second order terms, in the expansion are quite involved. Benaroya and Rehak (1987), analyzing a single-degree-of-freedom system and using the symbolic manipulation program MACSYMA, were not able to include more than the second term in the expansion. Yamazaki et.al (1985) suggested applying a Monte Carlo approach to equation (3.22) to simulate the processes $\alpha_k(\mathbf{x}, \theta)$ and hence the operator $\mathbf{\Pi}(\alpha(\mathbf{x}, \theta), \mathbf{x})$. At this stage, the successive orders in the expansion can be obtained numerically. Note that with this implementation sequence, only the deterministic average operator $\mathbf{L}^{-1}(\mathbf{x})$ needs to be inverted. However, the method still requires simulation, a fact that necessitates several runs of the simulated problem in order to assess the reliability of the results. Further, it can be seen that it is quite difficult to extend the method to obtain higher order moments than the first two. With regards to representing the random process, the same discussion presented for the perturbation method is pertinent. Namely, the processes are replaced by their local averages over the finite elements. In terms of implementation into a finite element code, either the variational or the Galerkin method may be used in equation (3.15). Specifically, replacing the random process by its local average over each finite element yields

$$[\, \mathbf{L} \, + \, \mathbf{\Pi} \,] \; \mathbf{U} \; = \; \mathbf{F} \,, \qquad (3.24)$$

where $\mathbf{L}$ is an $n \times n$ deterministic matrix, $\mathbf{\Pi}$ is an $n \times n$ random matrix function of the random variables $\{\alpha_i(\theta)\}_{i=1}^{r}$, and $\mathbf{F}$ and $\mathbf{U}$ are n-dimensional vectors representing the excitation and the response, respectively. Here, n denotes the number of degrees of freedom in the finite element mesh. At this point, a Neumann expansion of the inverse operator may be performed, leading to

$$\begin{aligned} \mathbf{U} \; &= \; [\, \mathbf{L} \, + \, \mathbf{\Pi} \,]^{-1} \; \mathbf{F} \\ &= \; \sum_{k=0}^{\infty} \left[\, \mathbf{L}^{-1} \, \mathbf{\Pi} \,\right]^{k} \; \mathbf{L}^{-1} \; \mathbf{F} \,. \end{aligned} \qquad (3.25)$$

Further, simulation can be used to numerically generate realizations of the matrix $\mathbf{\Pi}$, producing a statistical population for the response vector $\mathbf{U}$, from which various statistics can be obtained. Equation (3.25) is valid provided that

$$\| \, \mathbf{L}^{-1} \, \mathbf{\Pi} \, \| \; < \; 1 \qquad (3.26)$$

where $\|.\|$ denotes some norm in $\mathbf{R}^{n \times n} \times \mathbf{\Theta}$.

In the next section an improved Neumann expansion is presented in conjunction with a Galerkin based finite element method. The method makes it possible to compute the moments of the response in an explicit form. Thus, the simulation required by the method described in this section is circumvented.

3.3.5 Improved Neumann Expansion

In this section the Karhunen-Loeve expansion presented in Chapter II is implemented in the Galerkin formulation of the finite element method. The result is an explicit expansion for the response process. Let $\mathbf{D}$ denote a domain in $\mathbf{R}^n$. Here, n refers to the physical dimension of the problem, and $\mathbf{D}$ to the actual domain occupied by the object being investigated. Let $\mathbf{w}(\mathbf{x}, \theta)$, $x \in \mathbf{D}$, $\theta \in \Omega$ denote a vector of random properties of this domain. These properties are assumed to be the realization of a second order vector random process. Further, it is assumed that $\mathbf{w}(\mathbf{x}, \theta)$ is represented by its mean value $\bar{\mathbf{w}}(\mathbf{x})$ and its covariance matrix $[\, \mathbf{C}_{ww}(\, \mathbf{x}_1\, ,\, \mathbf{x}_2\,)\,]$. The ij^{th} component of this matrix represents the cross-correlation function between the processes $w_i(\mathbf{x}, \theta)$ and $w_j(\mathbf{x}, \theta)$ taken at $\mathbf{x} = \mathbf{x}_1$ and $\mathbf{x} = \mathbf{x}_2$ respectively. Let the domain $\mathbf{D}$ be subjected to a general external excitation denoted by $f(\mathbf{x}, \theta)$.

In general the response vector $u(\mathbf{x}, \theta)$ is related to the external excitation through a transformation defined by the medium $\mathbf{D}$ and its physical properties, resulting in equation (3.15) which is rewritten here for convenience

$$[\, \mathbf{L}(\mathbf{x}) \, + \, \mathbf{\Pi}(\, \alpha(\mathbf{x}, \theta)\, ,\, \mathbf{x}\,)\,]\, [u(\, \alpha(\mathbf{x}, \theta)\, ,\, \mathbf{x}\,)] \; = \; f(\, \mathbf{x}\, ,\, \theta\,). \qquad (3.27)$$

Further, let the domain be subjected to a set of boundary conditions prescribed by the operator equation

$$\mathbf{\Sigma}(\mathbf{x}, \theta)\, [u(\mathbf{x}, \theta)] \; = \; 0\, , \qquad \mathbf{x} \in \partial\mathbf{D}\, , \qquad (3.28)$$

where $\mathbf{\Sigma}$ is a random operator and $\partial\mathbf{D}$ denotes the boundary of the domain $\mathbf{D}$. It is obvious that $u(\mathbf{x}, \theta)$ belongs to the Hilbert space of functions satisfying equation (3.27) in $\mathbf{D}$ and such that equation (3.28) is satisfied on $\partial\mathbf{D}$. However, in view of the discussion of section (3.1), this space may be extended, through integration by parts to include functions not smooth enough to be operated upon as required by equation (3.27). To avoid obscuring the notation and to be able to proceed without digressing to a specific problem, it is assumed that only one parameter of the medium is random and that it appears as a multiplicative factor in the operator $\mathbf{\Pi}(\alpha(\mathbf{x}, \theta), \mathbf{x})$. Also it is assumed that the medium is subjected to a set of deterministic boundary conditions. Then, the covariance matrix $\mathbf{C}_{ww}(\mathbf{x}_1, \mathbf{x}_2)$ reduces to a single function $C_{ww}(\mathbf{x}_1, \mathbf{x}_2)$. In view of that, equations (3.27) and (3.28) become

$$[\, \mathbf{L}(\mathbf{x}) \, + \, \alpha(\mathbf{x}, \theta)\, \mathbf{R}(\mathbf{x})\,]\, [u(\mathbf{x}, \theta)] \; = \; f(\mathbf{x}, \theta) \qquad (3.29)$$

and

$$\mathbf{\Sigma}(\mathbf{x}) \, [u(\mathbf{x}, \theta)] \; = \; 0 \, , \tag{3.30}$$

where $\mathbf{R}(\mathbf{x})$ is a deterministic operator. Denote by $g_i(\mathbf{x})$ a basis for the admissible space corresponding to the operators $\mathbf{L}(\mathbf{x})$ and $\mathbf{R}(\mathbf{x})$, as discussed in section (3.1). Specifically, let $g_i(\mathbf{x})$ be the set of piecewise polynomials of degree high enough to allow it to span the admissible space. This choice for the basis induces a discretization of the domain $\mathbf{D}$ and makes the coefficients of the generalized coordinates $g_i(\mathbf{x})$ directly related to the components $u(\mathbf{x}, \theta)$ at the nodes of the induced mesh. The response can be written as

$$u(\mathbf{x} \, , \; \theta) \; = \; \sum_{i=1}^{N} \, u_i(\theta) \, g_i(\mathbf{x}) \, , \tag{3.31}$$

where $u_i(\theta)$ denotes the random magnitude of the ith degree of freedom. Substituting equation (3.31) into equation (3.29) and dropping the argument θ for convenience, gives

$$\sum_{i=1}^{N} \, u_i \, [\, \mathbf{L}(\mathbf{x}) \; + \; \alpha(\mathbf{x}) \, \mathbf{R}(\mathbf{x}) \,] \, g_i(\mathbf{x}) \; = \; f(\mathbf{x}) \, . \tag{3.32}$$

Multiplying equation (3.32) throughout by $g_i(\mathbf{x})$ and integrating over $\mathbf{D}$ yields

$$\sum_{i=1}^{N} \, \Big[\, \int_{\mathbf{D}} \, [\, \mathbf{L}(\mathbf{x}) \; + \alpha(\mathbf{x}) \, \mathbf{R}(\mathbf{x}) \,] \; \; g_i(\mathbf{x}) \, g_j(\mathbf{x}) \, d\mathbf{x} \,] \, u_i \tag{3.33}$$

$$= \; \int_{\mathbf{D}} \, f(\mathbf{x}) \, g_j(\mathbf{x}) \, d\mathbf{x} \, .$$

This equation can be rewritten as

$$\sum_{i=1}^{N} \, \Big[\, \int_{\mathbf{D}} \, [\, \mathbf{L}(\mathbf{x}) \, g_i(\mathbf{x}) \,] \; \, g_j(\mathbf{x}) d\mathbf{x} \tag{3.34}$$

$$+ \int_{\mathbf{D}} \, \alpha(\mathbf{x}) \, [\, \mathbf{R}(\mathbf{x}) \, g_i(\mathbf{x}) \,] \, g_j(\mathbf{x}) \, d\mathbf{x} \, \Big] \, u_i \; = \; \int_{\mathbf{D}} \, f(\mathbf{x}) \, g_j(\mathbf{x}) \, d\mathbf{x} \, .$$

Note that if, at this point, $\alpha(\mathbf{x}, \theta)$ is expanded along the same basis as $u(\mathbf{x}, \theta)$, in terms of its pointwise realization, the left hand-side of equation

(3.34) would involve a random matrix with fully correlated elements. This approach forms the basis for the methods discussed at the beginning of this chapter. Instead, $\alpha(\mathbf{x}, \theta)$ is expressed in its Karhunen-Loeve expansion, as described in section 3.3, in the form

$$\alpha(\mathbf{x}) = \sum_{n=1}^{M} \lambda_n \, \xi_n \, a_n(\mathbf{x}) \,, \tag{3.35}$$

where λ_n , $a_n(\mathbf{x})$ are the eigenvalue/eigenvector doublets corresponding to the covariance function of $\alpha(\mathbf{x}, \theta)$, and ξ_n is a set of orthonormal random variables. Substituting equation (3.35) for $\alpha(\mathbf{x}, \theta)$ into equation (3.34) gives

$$\sum_{i=1}^{N} \left[\bar{\mathbf{K}}_{ij} + \sum_{n=1}^{M} \xi_n \, \mathbf{K}_{ij}^{(n)} \right] \mathbf{u}_i = \mathbf{f}_j \,, \tag{3.36}$$

where

$$\bar{\mathbf{K}}_{ij} = \int_{\mathbf{D}} [\, \mathbf{L}(\mathbf{x}) \, g_i(\mathbf{x}) \,] \, g_j(\mathbf{x}) \, d\mathbf{x} \tag{3.37}$$

$$\mathbf{K}_{ij}^{(n)} = \int_{\mathbf{D}} a_n(\mathbf{x}) \, [\, \mathbf{L}(\mathbf{x}) \, g_i(\mathbf{x}) \,] \, g_j(\mathbf{x}) \, d\mathbf{x} \tag{3.38}$$

$$\mathbf{f}_j = \int_{\mathbf{D}} f(\mathbf{x}) \, g_j(\mathbf{x}) \, d\mathbf{x} \,. \tag{3.39}$$

Note that with the choice of the set $g_i(\mathbf{x})$ as piecewise polynomials, the integrals over $\mathbf{D}$ in the previous equations reduce to integrals over a small number of elements. Writing equation (3.36) for all values of $j = 1, \dots, N$, leads to a set of N equations to be solved for the random response at the various nodes,

$$\left[\bar{\mathbf{K}} + \sum_{n=1}^{M} \xi_n \, \mathbf{K}^{(n)} \right] \mathbf{u} = \mathbf{f} \tag{3.40}$$

where the matrices are defined by equations (3.37)-(3.39). At this stage the boundary conditions may be imposed on $\bar{\mathbf{K}}$ and each of the $\mathbf{K}^{(n)}$ matrices individually. A detailed account of this procedure is deferred to the next section where the problem is again encountered. Equation (3.40) can be normalized by multiplying throughout by $\bar{\mathbf{K}}^{-1}$ to obtain

$$\left[\mathbf{I} + \sum_{n=1}^{M} \xi_n \mathbf{Q}^{(n)} \right] \mathbf{u} = \mathbf{g} \qquad (3.41)$$

where

$$\mathbf{Q}^{(n)} = \bar{\mathbf{K}}^{-1} \mathbf{K}^{(n)} \qquad (3.42)$$

$$\mathbf{g} = \bar{\mathbf{K}}^{-1} \mathbf{f} . \qquad (3.43)$$

Equation (3.41), governing the behavior of the discretized system, can be formally written as

$$[\mathbf{I} + \boldsymbol{\Psi}[\{\xi_n\}]] \mathbf{u} = \mathbf{g} , \qquad (3.44)$$

where $\boldsymbol{\Psi}[\{\xi_n\}]$ is some functional of its arguments. Symbolically, equation (3.44) suggests that, in general, the response vector $\mathbf{u}$ is a nonlinearly filtered version of ξ_n. A straightforward scheme for obtaining the response vector from equation (3.44) relies on performing a Neumann expansion for the inverse operator to derive

$$\mathbf{u} = \left[\sum_{i=0}^{\infty} (-1)^i \boldsymbol{\Psi}^i[\{\xi_n\}] \right] \mathbf{g} . \qquad (3.45)$$

In terms of $\mathbf{Q}^{(n)}$, the response vector takes the form

$$\mathbf{u} = \sum_{i=0}^{\infty} (-1)^i \left[\sum_{n=1}^{M} \xi_n \mathbf{Q}^{(n)} \right]^i \mathbf{g} . \qquad (3.46)$$

Expanding equation (3.46) gives

$$\mathbf{u} = \left[\mathbf{I} - \sum_{n=1}^{M} \xi_n \mathbf{Q}^{(n)} + \sum_{m=1}^{M} \sum_{n=1}^{M} \xi_m \xi_n \mathbf{Q}^{(m)} \mathbf{Q}^{(n)} \right.$$
$$\left. - \sum_{k=1}^{M} \sum_{m=1}^{M} \sum_{n=1}^{M} \xi_k \xi_m \xi_n \mathbf{Q}^{(k)} \mathbf{Q}^{(m)} \mathbf{Q}^{(n)} + \dots \right] \mathbf{g}. \qquad (3.47)$$

Note that in the limit as $M \rightarrow \infty$, the results obtained by the preceding approach are identical with the results obtained from a direct Neumann expansion solution as described in section (3.3.3). The present formulation, however, provides an expression which is computationally tractable and amenable to automation.

3.3.6 Projection on the Homogeneous Chaos

The development in the previous section may be considered as a useful modification of existing techniques. Yet, it is limited in its applicability by equation (3.26). Clearly, this restriction is not prohibitively severe. However, given the range of applications encountered in practice it is desirable to devise a formulation which is more versatile. In addition, all the methods described previously lack the geometrical appeal that underlies the deterministic finite element method. In this section, a formulation of the problem exhibiting this attribute is introduced as a natural extension of the geometrical concepts of Hilbert spaces that form the basis for the deterministic finite element method. Two approaches are described that embody these concepts and lead to identical final equations.

Equations (3.29) and (3.30) constitute the starting point. Expanding $\alpha(\mathbf{x}, \theta)$ in a Karhunen-Loeve series gives

$$\left[\mathbf{L}(\mathbf{x}) + \sum_{n=1}^{M} \xi_n \, g_n(\mathbf{x}) \, \mathbf{R}(\mathbf{x}) \right] u(\mathbf{x}, \theta) = f(\mathbf{x}, \theta) . \qquad (3.48)$$

Assuming, without loss of generality, that $u(\mathbf{x}, \theta)$ is a second order process, it lends itself to a Karhunen-Loeve expansion of the form

$$u(\mathbf{x}, \theta) = \sum_{j=1}^{L} e_j \, \chi_j(\theta) \, b_j(\mathbf{x}) , \qquad (3.49)$$

where

$$\int_{\mathbf{D}} C_{uu} (\mathbf{x_1} , \mathbf{x_2}) \, b_j(\mathbf{x_2}) \, d\mathbf{x_2} = e_j \, b_j(\mathbf{x_1}) , \qquad (3.50)$$

and

$$\chi_j(\theta) = \frac{1}{e_j} \int_{\mathbf{D}} u(\mathbf{x}, \theta) \, b_j(\mathbf{x}) \, d\mathbf{x} . \qquad (3.51)$$

Obviously, the covariance function $C_{uu}(\mathbf{x}_1, \mathbf{x}_2)$ of the response process is not known at this stage. Thus, e_j and $b_j(\mathbf{x})$ are also not known. Further, $u(\mathbf{x}, \theta)$, not being a Gaussian process, the set $\chi_j(\theta)$ is not a Gaussian vector. Therefore, equation (3.48) is of little use in its present form. Relying on the discussion of section (2.4) concerning the Homogeneous Chaos, the second order random variables $\chi_j(\theta)$ can be represented by the mean-square convergent expansion

$$
\chi_j(\theta) \;=\; a_{i_0}^{(j)}\, \Gamma_0 \;+\; \sum_{i_1=1}^{\infty} a_{i_1}^{(j)}\, \Gamma_1(\xi_{i_1})
$$

$$
+\; \sum_{i_1=1}^{\infty} \sum_{i_2=1}^{i_1} a_{i_1,i_2}^{(j)}\, \Gamma_2(\xi_{i_1}, \xi_{i_2}) + \sum_{i_1=1}^{\infty} \sum_{i_2=1}^{i_1} \sum_{i_3=1}^{i_2} a_{i_1 i_2 i_3}^{(j)}\, \Gamma_3(\xi_{i_1}, \xi_{i_2}, \xi_{i_3})
$$

$$
+\; \sum_{i_1=1}^{\infty} \sum_{i_2=1}^{i_1} \sum_{i_3=1}^{i_2} \sum_{i_4=1}^{i_3} a_{i_1 i_2 i_3 i_4}^{(j)}\, \Gamma_4(\xi_{i_1},\ \xi_{i_2},\ \xi_{i_3},\ \xi_{i_4}) + \dots, \qquad (3.52)
$$

where $a_{i_1,\dots i_p}^{(j)}$ are deterministic constants independent of θ and $\Gamma_p(\xi_{i_1},\ \dots,\ \xi_{i_p})$ is the pth order Homogeneous Chaos. Equation (3.52) is truncated after the Pth polynomial and is rewritten for convenience, as discussed in equation (2.86), in the following form,

$$
\chi_j(\theta) \;=\; \sum_{i=0}^{P} x_i^{(j)}\, \Psi_i[\{\xi_r\}]\,, \qquad (3.53)
$$

where $x_i^{(j)}$ and $\Psi_i[\{\xi_r\}]$ are identical to $a_{i_1\dots i_p}^{(j)}$ and $\Gamma_p(\xi_{i_1},\dots\xi_{i_p})$, respectively. In equation (3.53), P denotes the total number of Polynomial Chaoses used in the expansion, excluding the $zero^{th}$ order term. Given the number M of terms used in the Karhunen-Loeve expansion, and the order p of Homogeneous Chaos used, P may be determined by the equation

$$
P \;=\; 1 \;+\; \sum_{s=1}^{p} \frac{1}{s!} \prod_{r=0}^{s-1} (M+r)\,. \qquad (3.54)
$$

Table 3.1 shows values of P for combinations of p and M

Substituting equation (3.53) for $\chi_k(\theta)$, equation (3.49) becomes

$$
u(\mathbf{x}, \theta) \;=\; \sum_{j=1}^{L} \sum_{i=0}^{P} x_i^{(j)}\, \Psi_i[\{\xi_r\}]\, c_j(\mathbf{x})\,, \qquad (3.55)
$$

M	\multicolumn{5}{c}{p=order of the Homogeneous Chaos}
	0	1	2	3	4
2	1	3	6	10	15
4	1	5	15	35	70
6	1	7	28	83	210

Table 3.1: Coefficient P for the size of the extended system
corresponding to selected values of M and p
M = order of the Karhunen-Loeve expansion.
p = order of the Homogeneous Chaos expansion.

where

$$c_j(\mathbf{x}) = e_j \, b_j(\mathbf{x}) \, . \tag{3.56}$$

Changing the order of summation in equation (3.55) gives

$$u(\mathbf{x}, \theta) = \sum_{i=0}^{P} \Psi_i[\{\xi_r\}] \sum_{j=1}^{L} x_i^{(j)} \, c_j(\mathbf{x})$$

$$= \sum_{i=0}^{P} \Psi_i[\{\xi_r\}] \, d_i(\mathbf{x}) \, , \tag{3.57}$$

where,

$$d_i(\mathbf{x}) = \sum_{k=1}^{L} x_i^{(j)} \, c_j(\mathbf{x}) \, . \tag{3.58}$$

Substituting equation (3.57) for $u(\mathbf{x}, \theta)$, equation (3.48) becomes

$$\left[\mathbf{L}(\mathbf{x}) + \sum_{n=1}^{M} \xi_n \, a_n(\mathbf{x}) \, \mathbf{R}(\mathbf{x}) \right] \sum_{j=0}^{P} \Psi_j[\{\xi_r\}] \, d_j(\mathbf{x}) = f(\mathbf{x}) \, , \tag{3.59}$$

where reference to the parameter θ was eliminated for notational simplicity.
The response $u(\mathbf{x}, \theta)$ can be completely determined once the functions $d_i(\mathbf{x})$
are known. In terms of the eigenfunctions $b_j(\mathbf{x})$ of the covariance function
of $u(\mathbf{x}, \theta)$, $d_i(\mathbf{x})$ can be expressed as

$$d_i(\mathbf{x}) = \sum_{j=1}^{L} x_i^{(j)} \, e_j \, b_j(\mathbf{x})$$

$$= \sum_{j=1}^{L} y_i^{(j)} \, b_j(\mathbf{x}) \ . \tag{3.60}$$

Note that, if $u(\mathbf{x}, \theta)$ satisfies homogeneous deterministic boundary conditions on some section of the boundary $\partial \mathbf{D}_1 \in \partial \mathbf{D}$, then $C_{uu}(\mathbf{x}_1, \mathbf{x}_2) \equiv 0$ for $\mathbf{x}_1 \in \partial \mathbf{D}_1$ or $\mathbf{x}_2 \in \partial \mathbf{D}_1$. Thus, according to equation (3.50), $b_j(\mathbf{x}) \equiv 0$ for $\mathbf{x} \in \partial \mathbf{D}_1$. It can then be deduced that the set $b_k(\mathbf{x})$ satisfies the homogeneous boundary conditions imposed on the problem and so do the elements of the set $d_i(\mathbf{x})$ by virtue of equation (3.60). Equation (3.59) may be written in a more suggestive form

$$\sum_{j=0}^{P} \Psi_j[\{\xi_j\}] \, \mathbf{L}(\mathbf{x}) \, d_j(\mathbf{x}) + \sum_{j=0}^{P} \sum_{i=1}^{M} \xi_i \, \Psi_j[\{\xi_r\}] \, \mathbf{R}(\mathbf{x}) \, d_j(\mathbf{x}) = f(\mathbf{x}). \tag{3.61}$$

This form of the equation shows that $d_j(\mathbf{x})$ belongs to the intersection of the domains of $\mathbf{R}(\mathbf{x})$ and $\mathbf{L}(\mathbf{x})$. Then, in view of the discussion of the deterministic finite element method in section (3.2), the function $d_j(\mathbf{x})$ may be expanded in the space $\mathbf{C}^m$ as

$$d_j(\mathbf{x}) = \sum_{k=1}^{N} d_{kj} \, g_k(\mathbf{x}) \ . \tag{3.62}$$

Then, equation (3.61) becomes

$$\sum_{j=0}^{P} \sum_{k=1}^{N} d_{kj} \, \Psi_j[\{\xi_r\}] \, \mathbf{L}(\mathbf{x}) \, g_k(\mathbf{x}) \tag{3.63}$$

$$+ \sum_{j=0}^{P} \sum_{i=1}^{M} \xi_i(\theta) \, \Psi_j[\{\xi_r\}] \sum_{k=1}^{N} d_{kj} \, \mathbf{R}(\mathbf{x}) \, g_k(\mathbf{x}) = f(\mathbf{x}) \ .$$

Equation (3.63) may be rearranged to give

$$\sum_{j=0}^{P} \sum_{k=1}^{M} d_{kj} \left[\, \Psi_j[\{\xi_r\}] \, \mathbf{L}(\mathbf{x}) \, g_k(\mathbf{x}) \right.$$

$$\left. + \sum_{i=1}^{M} \xi_i(\theta) \, \Psi_j[\{\xi_r\}] \, \mathbf{R}(\mathbf{x}) \, g_k(\mathbf{x}) \right] = f(\mathbf{x}) \ . \tag{3.64}$$

Multiplying both sides of equation (3.64) by $g_l(\mathbf{x})$ and integrating throughout yields

$$
\sum_{j=0}^{P} \sum_{k=1}^{M} d_{kj} \left[\Psi_j[\{\xi_r\}] \int_{\mathbf{D}} \mathbf{L}(\mathbf{x}) \, g_k(\mathbf{x}) \, g_l(\mathbf{x}) \, d\mathbf{x} \right.
$$

$$
+ \sum_{i=1}^{M} \xi_i \, \Psi_j[\{\xi_r\}] \left. \int_{\mathbf{D}} \mathbf{R}(\mathbf{x}) \, g_k(\mathbf{x}) \, g_l(\mathbf{x}) \, d\mathbf{x} \right]
$$

$$
= \int_{\mathbf{D}} f(\mathbf{x}) \, g_l(\mathbf{x}) \, d\mathbf{x} \, , \ l \, = \, 1 \,, ..., \, N \, . \tag{3.65}
$$

Setting

$$
\mathbf{L}_{kl} \; = \; \int_{\mathbf{D}} \mathbf{L}(\mathbf{x}) \, g_k(\mathbf{x}) \, g_l(\mathbf{x}) \, d\mathbf{x} \tag{3.66}
$$

$$
\mathbf{R}_{ikl} \; = \; \int_{\mathbf{D}} \mathbf{R}(\mathbf{x}) \, g_k(\mathbf{x}) \, g_l(\mathbf{x}) \, a_i(\mathbf{x}) \, d\mathbf{x} \tag{3.67}
$$

$$
\mathbf{f}_l \; = \; \int_{\mathbf{D}} f(\mathbf{x}) \, g_l(\mathbf{x}) \, d\mathbf{x} \, , \tag{3.68}
$$

equation (3.65) becomes

$$
\sum_{j=0}^{P} \sum_{k=1}^{N} \left[\Psi_j[\{\xi_r\}] \, \mathbf{L}_{kl} \, + \, \sum_{i=1}^{M} \xi_i(\theta) \, \Psi_j[\{\xi_r\}] \, \mathbf{R}_{ikl} \right] d_{kj}
$$

$$
= \; \mathbf{f}_l \, , \, l = 1, ..., N \, . \tag{3.69}
$$

Note that the index j spans the number of Polynomial Chaoses used, while the index k spans the number of basis vectors used in $\mathbf{C}^m$. Multiplying equation (3.69) by $\Psi_m[\{\xi_r\}]$, averaging throughout and noting that

$$
<\Psi_j[\{\xi_r\}] \, \Psi_m[\{\xi_r\}]> \; = \; \delta_{jm} <\Psi_m^2[\{\xi_r\}]> \, , \tag{3.70}
$$

one can derive

$$
\sum_{k=1}^{N} <\Psi_m^2[\{\xi_r\}]>\mathbf{L}_{kl} d_{km} + \sum_{j=0}^{P} \sum_{k=1}^{N} d_{kj} \sum_{i=1}^{M} <\xi_i(\theta)\Psi_j[\{\xi_r\}]\Psi_m[\{\xi_r\}]>\mathbf{R}_{ikl}
$$

$$
= <\mathbf{f}_l \, \Psi_m[\{\xi_r\}]>, \quad l \, = \, 1, ..., N \, , \, m \, = \, 1, ..., M \, . \tag{3.71}
$$

i	Indices of the Polynomial Chaoses	$c_{ijk}^{(2)}$
	j	k	

i	j	k	$c_{ijk}^{(2)}$
i=1	1	3	2
	2	4	1
	3	6	6
	4	7	2
	5	8	2
	6	10	24
	7	11	6
	8	12	4
	9	13	6
i=2	1	4	1
	2	5	2
	3	7	2
	4	8	2
	5	9	6
	6	11	6
	7	12	4
	8	13	6
	9	14	24

Table 3.2: Coefficient $c_{ijk}^{(2)} = <\xi_i \Psi_j \Psi_k>$, $c_{ijk}^{(2)} = c_{ikj}^{(2)}$ Two-Dimensional Polynomial Chaoses; equation (3.72).

Introducing

$$c_{ijm} \equiv <\xi_i \Psi_j[\{\xi_r\}] \Psi_m[\{\xi_r\}]> , \qquad (3.72)$$

and assuming, without loss of generality, that the Polynomial Chaoses have been normalized, equation (3.71) becomes

$$\sum_{k=1}^{N} \mathbf{L}_{kl} d_{km} + \sum_{j=0}^{P} \sum_{k=1}^{N} d_{kj} \sum_{i=1}^{M} \mathbf{R}_{ikl} c_{ijm} = <\mathbf{f}_l \Psi_m[\{\xi_r\}]> ,$$
$$l = 1, ..., N , m = 1, ..., M . \qquad (3.73)$$

For a large number of index combinations the coefficients c_{ijm} are identically zero. Equation (3.72) was implemented using the symbolic manipulation program MACSYMA (1986). The non-zero values corresponding to

i	Indices of the Polynomial Chaoses		$c_{ijk}^{(2)}$
	j	k	
i=1	1	5	2
	2	6	1
	3	7	1
	4	8	1
	5	15	6
	6	16	2
	7	17	2
	8	18	2
	9	19	2
	10	20	1
	11	21	1
	12	22	2
	13	23	1
	14	24	2
	15	35	24
	16	36	6
	17	37	6
	18	38	6
	19	39	4
	20	40	2
	21	41	2
	22	42	4
	23	43	2
	24	44	4
	25	45	6
	26	46	2
	27	47	2
	28	48	2
	29	49	1
	30	50	2
	31	51	6
	32	52	2
	33	53	2
	34	54	6

Table 3.3: Coefficient $c_{ijk}^{(4)} = <\xi_i \ \Psi_j \ \Psi_k>$, $c_{ijk}^{(4)} = c_{ikj}^{(4)}$ Four-Dimensional Polynomial Chaoses; equation (3.72).

i	Indices of the Polynomial Chaoses		$c_{ijk}^{(4)}$
	j	k	
i=2	1	6	1
	2	9	2
	3	10	1
	4	11	1
	5	16	2
	6	19	2
	7	20	1
	8	21	1
	9	25	6
	10	26	2
	11	27	2
	12	28	2
	13	29	1
	14	30	2
	15	36	6
	16	39	4
	17	40	2
	18	41	2
	19	45	6
	20	46	2
	21	47	2
	22	48	2
	23	49	1
	24	50	2
	25	55	24
	26	56	6
	27	57	6
	28	58	4
	29	59	2
	30	60	4
	31	61	6
	32	62	2
	33	63	2
	34	64	6

Table 3.3 (continued): Coefficient $c_{ijk}^{(4)} = <\xi_i \ \Psi_j \ \Psi_k>$, $c_{ijk}^{(4)} = c_{ikj}^{(4)}$ Four-Dimensional Polynomial Chaoses; equation (3.72).

i	Indices of the Polynomial Chaoses		$c_{ijk}^{(4)}$
	j	k	
i=3	1	7	11
	2	10	1
	3	12	2
	4	13	1
	5	17	2
	6	20	1
	7	22	2
	8	23	1
	9	26	2
	10	28	2
	11	29	1
	12	31	6
	13	32	2
	14	33	2
	15	37	6
	16	40	2
	17	42	4
	18	43	2
	19	46	2
	20	48	2
	21	49	1
	22	51	6
	23	52	2
	24	53	2
	25	56	6
	26	58	4
	27	59	2
	28	61	6
	29	62	2
	30	63	2
	31	65	24
	32	66	6
	33	67	4
	34	68	6

Table 3.3 (continued): Coefficient $c_{ijk}^{(4)} = <\xi_i \ \Psi_j \ \Psi_k>$, $c_{ijk}^{(4)} = c_{ikj}^{(4)}$ Four-Dimensional Polynomial Chaoses; equation (3.72).

i	Indices of the Polynomial Chaoses		$c_{ijk}^{(4)}$
	j	k	
i=4	1	8	1
	2	11	1
	3	13	1
	4	14	2
	5	18	2
	6	21	1
	7	23	1
	8	24	2
	9	27	2
	10	29	1
	11	30	2
	12	32	2
	13	33	2
	14	34	6
	15	38	6
	16	41	2
	17	43	2
	18	44	4
	19	47	2
	20	49	1
	21	50	2
	22	52	2
	23	53	2
	24	54	6
	25	57	6
	26	59	2
	27	60	4
	28	62	2
	29	63	2
	30	64	6
	31	66	6
	32	67	4
	33	68	6
	34	69	24

Table 3.3 (continued): Coefficient $c_{ijk}^{(4)} = <\xi_i \ \Psi_j \ \Psi_k>$, $c_{ijk}^{(4)} = c_{ikj}^{(4)}$ Four-Dimensional Polynomial Chaoses; equation (3.72).

the first four Homogeneous Chaoses are shown in Tables 3.2 and 3.3 for two and four terms in the Karhunen-Loeve expansion ($K = 2$ and $K = 4$), respectively. Forming equation (3.73) for all P values of m, produces a set of $N \times M$ algebraic equations of the form

$$[\, \mathbf{G} + \mathbf{R} \,]\ \mathbf{d}\ =\ \mathbf{h}\,, \tag{3.74}$$

where $\mathbf{G}$ and $\mathbf{R}$ are block matrices of dimension $N \times M$. Their mj^{th} blocks are N-dimensional square matrices given by the equations

$$\mathbf{G}_{mj}\ =\ \delta_{mj}\,\mathbf{L}\,, \tag{3.75}$$

and

$$\mathbf{R}_{mj}\ =\ \sum_{i=1}^{M} c_{ijm}\,\mathbf{R}_i\,. \tag{3.76}$$

In equations (3.75) and (3.76), $\mathbf{L}$ and $\mathbf{R}_i$ denote N-dimensional square matrices whose kl^{th} element is given by equations (3.66) and (3.67), respectively. In equation (3.74), $\mathbf{h}$ signifies the $N \times M$ vector whose m^{th} block is given by the equation

$$\mathbf{h}_m\ =\ <\mathbf{f}\ \Psi_m[\{\xi_r\}]>\,. \tag{3.77}$$

The N-dimensional vectors $\mathbf{d}_m$ can be obtained as the subvectors of the solution to the deterministic algebraic problem given by equation (3.74). Once these coefficients are obtained, back substituting into equation (3.57) yields an expression of the response process in terms of the Polynomial Chaoses of the form

$$\mathbf{u}\ =\ \sum_{j=0}^{P}\ \mathbf{d}_j\ \Psi_j[\{\xi_r\}]\,. \tag{3.78}$$

Equation (3.74) can also be derived by a different approach, which is a simple variation of what has been presented above. In fact, if the computations leading to equation (3.41) are carried out in the same manner as above, an equation is obtained involving a set of unknown random variables representing the response at the nodal points of the finite element mesh. For convenience equation (3.41) is rewritten as

$$\left[\mathbf{I}\ +\ \sum_{k=1}^{M} \xi_k(\theta)\ \mathbf{Q}^{(k)}\right]\ \mathbf{u}\ =\ \mathbf{g}\,. \tag{3.79}$$

Further, each element of the vector $\mathbf{u}$, being a second-order random variable, can be expanded as in equation (3.53) to obtain

$$u_i = \sum_{j=0}^{P} c_{ij}\ \Psi_j[\{\xi_r\}]\ , \qquad (3.80)$$

or,

$$\mathbf{u} = \sum_{j=0}^{P} \mathbf{c}_j\ \Psi_j[\{\xi_r\}]\ , \qquad (3.81)$$

where $\mathbf{c}_j$ is a vector of the same dimension as $\mathbf{u}$. Substituting equation (3.81) back into equation (3.79) gives

$$\sum_{j=0}^{P} \left[\mathbf{I} + \sum_{i=1}^{M} \xi_i(\theta)\ \mathbf{Q}^{(i)} \right] \mathbf{c}_j\ \Psi_j[\{\xi_r\}] = \mathbf{g}\ . \qquad (3.82)$$

Forming the inner product over $\boldsymbol{\Theta}$ of equation (3.81) with $\{\Psi_m[\{\xi_r\}]\}_{m=1}^{P}$, gives

$$\mathbf{c}_m + \sum_{j=0}^{P} \sum_{i=1}^{M} c_{ijm}\ \mathbf{Q}^{(i)}\ \mathbf{c}_j = <\mathbf{g}\ \Psi_m[\{\xi_r\}]>\ . \qquad (3.83)$$

Carrying out the P indicated inner products yields an equation identical to equation (3.74) with $\mathbf{G} \equiv \mathbf{I}$.

3.3.7   Geometrical and Variational Extensions

Geometry of the Solution Process

The development in the last section hinges on the fact that the Polynomial Chaoses form a complete basis in the Hilbert space $\boldsymbol{\Theta}$, or equivalently that the Homogeneous Chaos ring $\boldsymbol{\Phi}_{\theta(\xi)}$ defined in Chapter I is dense in $\boldsymbol{\Theta}$. The orthogonality property of the Polynomial Chaoses is not a requirement for the Galerkin procedure and is the result of the construction procedure followed in section (2.4.3). Unlike the deterministic case, where the set $g_k(\mathbf{x})$ has to satisfy certain smoothness and boundary conditions, any complete basis in $\boldsymbol{\Theta}$ can be used as an alternative to the Polynomial Chaoses in the development of the stochastic finite element method. Obviously, one such set is the set of the homogeneous polynomials formed by simple products of the elements of the set ξ_r. These polynomials are not orthogonal but they are easier to generate than the Polynomial Chaoses. Using such a set in the previous development would yield an expansion of the response process $u(\mathbf{x}, \theta)$ in the form

$$u(\mathbf{x}, \theta) = \sum_{k=0}^{\infty} \sum_{\rho_1 + \ldots + \rho_r \leq k} a_{i_1 \ldots i_r}^{\rho_1 \ldots \rho_r} \, \xi_{i_1}^{\rho_1} \, \ldots \, \xi_{i_r}^{\rho_r} \, . \tag{3.84}$$

Equation (3.84) is similar to equation (3.47) obtained in section (3.3.4) using the improved Neumann expansion. In the Neumann expansion case, however, the optimality criterion, in terms of some norm of the error, is not well defined. The Neumann expansion can then be viewed as a special case of a more general procedure, whereby the response process is expressed in terms of a complete set in the space Θ of second-order random variables. Viewed from this geometrical perspective, the methods presented in this Chapter lend themselves to interesting interpretations. Namely, the perturbation approach is a Taylor series expansion, the validity and convergence of which depend upon the behavior of the response process in the neighborhood of its mean value. The Neumann expansion is an expression of the response in terms of a complete non-orthogonal basis in the space Θ. Finally, the Homogeneous Chaos approach is an orthogonal expansion in Θ of the response process. Clearly, each of these methods is based on a different error minimization criterion. Unlike the deterministic case where non-orthogonal bases may have the merit of being flexible so as to satisfy the smoothness and boundary conditions, there is no merit for using a non-orthogonal expansion in Θ, since such conditions do not exist. In fact, as will be observed in the next chapter by comparing the resulting expressions for the covariance matrices of the response, orthogonal expansion yields much simpler forms for the response statistics. Also it is noted in Chapter V that the Homogeneous Chaos approach has better convergence properties than the Neumann expansion approach. In this context, it is worth pointing out the similarity of the approximation scheme employed here to the p-method of the deterministic finite element method discussed earlier. Indeed, the interval along the random dimension is not discretized, and a single element is used in the approximation. Instead, the refinement in the results is achieved by increasing the order of the functional polynomial approximation.

Variational Extensions

It may be observed that the development of the Homogeneous Chaos expansion may be considered as an extension of the Galerkin finite element method to the stochastic case. Indeed, the error resulting from a finite expansion is made orthogonal, in the sense of the inner product in Θ, to the subspace spanned by the approximating basis. A question that arises in this context involves the possibility of developing an analogous theory based on the extension of the variational principles to the stochastic case. The variational approach in the deterministic case usually expresses the constraint that the true solution to a given problem is that function which minimizes a certain norm or measure of the energy released by the system. The stochastic parallel to this energy norm would be a measure of the stochastic energy dissipated by the system. Stochastic energy may be interpreted as being the uncertainty associated with the system under consideration. Such an uncertainty is well known in the context of information theory and it has an associated measure. That is the information-entropy of the system. Indeed, a related variational principle can be formulated based on the concept of maximum entropy. It may then seem plausible to develop a stochastic variational principle based on the concepts presented above. At this point, however, there is no indication as to how to assemble these ideas into a coherent theory.

Chapter 4

STOCHASTIC FINITE ELEMENT METHOD: Response Statistics

4.1 Reliability Theory Background

An important objective of a stochastic finite element analysis of an engineering system should be the determination of a set of design criteria which can be implemented in a probabilistic context. In other words, it is important that any method that is used for the numerical treatment of stochastic systems be compatible with some rationale that permits a reliability analysis. Traditionally, this problem has been addressed by relying on the second order statistical moments of the response process. These moments offer useful statistical information regarding the response of the system and equations will be developed in an ensuing section for their determination. However, second order moments do not suffice for a complete reliability analysis. For this, higher order moments and related information are required. In this context, the response of any system to a certain excitation may be viewed as a point in the space defined by the parameters describing the system. In other words, for every set of such parameters, the response is uniquely determined as a function of the independent variables. When these parameters are regarded as random variables, the response function may be described as a point in the space spanned by these variables, defining a surface that corresponds to all the possible realizations of the random parameters. The case where the parameters are modeled as random processes may be reduced, as discussed in Chapter II, to the case of random variables. For the safe operation of a given system, it is usually required that the response, or some measure of it, be confined to a certain region in this space, termed the safe region. Formally, this condition can be expressed as

$$g[\{\xi_i\}_{i=1}^r] \geq 0 \ , \tag{4.1}$$

where

$$g[\{\xi_i\}_{i=1}^r] = 0 \tag{4.2}$$

is the equation of the failure surface in the $\boldsymbol{\xi}$-space. Of special interest in reliability assessment is the failure probability, P_f. That is, the probability that equation (4.1) is not satisfied. This equals the volume, with respect to some probability measure, of the region in the $\boldsymbol{\xi}$-space over which (4.1) is not satisfied. That is,

$$P_f = \int_{g[\{\xi_i\}]\geq 0} f_{\boldsymbol{\xi}}(\boldsymbol{\xi}) \, d\boldsymbol{\xi} \ , \tag{4.3}$$

where $f_{\boldsymbol{\xi}}(\boldsymbol{\xi})$ is the joint probability distribution of $\{\xi_1, ..., \xi_r\}$. The difficulty in using equation (4.3) in engineering practice is obvious. Specifically, the domain of integration is rarely available in an analytical form. Further, the dimension of the ξ-space is usually quite large, to the extent that the requisite integration can seldom be carried out either analytically or numerically (Schueller and Stix, 1987). Finally, realizations of the set $\{\xi_i\}$ are often scarce and do not permit inference about their joint distribution beyond their joint second order moments. Grigoriu et.al (1979) have suggested treating the selection of this joint distribution as a problem of decision theory and proposed a number of optimality criteria to resolve this issue. The problem remains, however, of efficiently implementing these distributions into numerical algorithms. This is a matter that persistently poses severe limitations on the amount of usable probabilistic information.

Given these computational and theoretical difficulties, it became desirable to introduce a standard measure for reliability analysis. Such a measure would be based on readily available second moment properties of the random variables involved and would reflect a safety measure associated with the operation of the system. The reliability index was introduced as an attempt to meet these needs. The basic idea behind the reliability index concept is the fact that the less uncertainty there is concerning the limit state of a system, the safer the system is. In other words, the greater the mean of the limit state relative to its standard deviation, the less likely is the response to wander beyond this state. As an intuitive measure, the reliability index was first defined as (Freudenthal, 1947; Freudenthal et. al, 1966; Cornell, 1969; Ditlevsen, 1981; Thoft-Christensen, 1982, 1983; Melchers, 1987; Madsen et.al, 1986)

$$\beta = \frac{<M>}{\sigma_M} , \qquad (4.4)$$

where M denotes the safety margin which is the difference between resistance and load effects, $<M>$ denotes its expected value, and σ_M its standard deviation. Note that, if the random variables $\{\xi_i\}$ are Gaussian and the limit state function given by equation (4.1) is linear in its arguments, the safety margin would have a Gaussian distribution. Further, β, as defined by equation (4.4), would be equal to the distance from the mean value of M to the hyperplane representing the limit state. For nonlinear limit states, the standard deviation of the safety margin is related to higher order moments of the system random variables which often are not known. In order to be able to use β, as given by equation (4.4), the limit state surface is usually linearized using the first term in a Taylor series about some point in the ξ-space. In this context, it is well known (Ditlevsen, 1981) that the mean value $<M>$ does not represent the best linearization point. Also note that if a function $g[.]$ satisfies equation (4.1), so does any function that is a power of $g[.]$, as well as any function that shares the same zeros with $g[.]$. This so called lack of invariance problem was resolved by introducing a new reliability index (Hasofer and Lind, 1974) defined by the equation

$$\beta_{HL} \equiv \min_{\mathbf{x} \in L_{\mathbf{x}}} \sqrt{(\mathbf{x} - <\boldsymbol{\xi}>)^T \, \mathbf{C}_{\boldsymbol{\xi}}^{-1} \, (\mathbf{x} - <\boldsymbol{\xi}>)} , \qquad (4.5)$$

where $L_{\mathbf{x}}$ denotes the limit surface and $\mathbf{C}_{\boldsymbol{\xi}}$ is the covariance matrix of the design variables $\{\xi_i\}$. Given the set $\{\xi_i\}$, β_{HL} is uniquely determined as the shortest distance to the limit state surface. Thus, it is seen that the optimal point for the linearization of the limit state function, is the point on that surface that is closest to the mean value after the design variables have been transformed to a set of uncorrelated variables. A deficiency of the Hasofer-Lind index is a lack of comparativeness (Ditlevsen, 1981) associated with the fact that any surface that is tangent to the failure surface at the design point has the same β_{HL}. Transforming the design variables to a new set having the Gaussian distribution (Rosenblatt, 1952) resolves this issue. When the failure surface is a hyperplane, manifested by the linearity of equation (4.1), the Hasofer-Lind index coincides with the previously defined index given by equation (4.4). The design point obtained in this way can be shown to be the point of maximum failure likelihood when the design variables are uncorrelated and Gaussian (Shinozuka, 1983). Given the reliability index, for Gaussian variables, the failure probability may be approximated by the expression

$$P_f \approx \Phi(-\beta_{HL}) , \qquad (4.6)$$

where Φ is the normal probability function. Equation (4.6) is exact for linear limit states and becomes approximate for nonlinear limit states. The main advantage of the first order second moment method as reviewed herein is its simplicity and ease of implementation. However, the method has some serious shortcomings, (Schueller and Stix, 1987), associated with the replacement of the failure surface by a tangent hyperplane; the magnitude of the error is hard to estimate but is known to increase drastically with increasing number of design variables.

4.2 Statistical Moments

4.2.1 Moments and Cummulants Equations

From the preceding review it becomes clear that the statistical moments of the response not only can provide bounds about its expected range, but they can also provide meaningful information about the reliability of the random system. Thus, it becomes desirable to develop equations for the determination of the statistical moments, up to an arbitrary order, of the response process. In this context the general series representation of the response, associated with the stochastic finite element method presented in Chapter III, is quite well suited for an efficient computation of the corresponding statistical moments.

As obtained either from the improved Neumann expansion or from the Homogeneous Chaos formulation, the response process vector may be represented by the series

$$\mathbf{u} \;=\; \mathbf{a}_0 \,+\, \mathbf{a}_{\lambda_1} \, \Pi_{\lambda_1}(\boldsymbol{\xi}) \,+\, \mathbf{a}_{\lambda_1\lambda_2} \, \Pi_{\lambda_1\lambda_2}(\boldsymbol{\xi}) \,+\, \cdots$$

$$= \sum_{k=0}^{\infty} \mathbf{a}_{\lambda_0 \ldots \lambda_k} \, \Pi_{\lambda_0 \ldots \lambda_k}(\boldsymbol{\xi}) \, , \tag{4.7}$$

where $\Pi_{\lambda_0 \ldots \lambda_k}(\boldsymbol{\xi})$ is a kth order polynomial in the variables $\{\xi_{\lambda_0}, \ldots, \xi_{\lambda_k}\}$ and $\mathbf{a}_{\lambda_0 \ldots \lambda_k}$ are deterministic vectors. Specifically, in the improved Neumann expansion, the polynomials Π are simply the homogeneous polynomials given by the equation

$$\Pi_{\lambda_0 \ldots \lambda_k}(\boldsymbol{\xi}) = \prod_{i=1}^{k} \xi_{\lambda_i} \, , \tag{4.8}$$

whereas for the Homogeneous Chaos formulation, these are the Polynomials Chaoses

$$\Pi_{\lambda_0 \ldots \lambda_k}(\boldsymbol{\xi}) = \Gamma_{\lambda_0 \ldots \lambda_k}(\xi_{\lambda_0}, \ldots, \xi_{\lambda_k}) \, . \tag{4.9}$$

In equation (4.7), indicial notation is used, and a repeated index implies summation with respect to that index over its range. McCullagh (1984, 1987) describes methods to obtain the cummulant generating function for a polynomial of the form given by equation (4.8). The rth component of $\mathbf{u}$, $u^{(r)}$ can be written as

$$u^{(r)} = P^{(r)} \, \boldsymbol{\xi} \tag{4.10}$$

where $P^{(r)}$ is the operator indicated by equation (4.7) as applicable to $u^{(r)}$. Then, following McCullagh (1987), the cummulant generating function for $u^{(r)}$ is given by the equation

$$K_u(\chi) = e^{\chi_r P^{(r)}} \, \kappa_{\boldsymbol{\xi}} \, , \tag{4.11}$$

where $\kappa_{\boldsymbol{\xi}}$ represents cummulants of $\boldsymbol{\xi}$, and $e^{\chi_r P^{(r)}}$ is an operator yielding upon expanding,

$$e^{\chi_r P^{(r)}} = \left[I + \chi_r \, P^{(r)} + \frac{1}{2} \, \chi_r \chi_s \, P^{(r)} P^{(s)} + \ldots \right] \, . \tag{4.12}$$

The result of the operation of $e^{\chi_r P^{(r)}}$ on $\kappa_{\boldsymbol{\xi}}$ is obtained by compounding the action of the individual components appearing in equation (4.12). Specifically, it is noted that

$$I \, \kappa_{\boldsymbol{\xi}} = 0 \tag{4.13}$$

$$a_i^{(r)} \, \kappa_{\boldsymbol{\xi}} \; = \; a_i^{(r)} \, \kappa^i \tag{4.14}$$

$$a_{ij}^{(r)} \, \kappa_{\boldsymbol{\xi}} \; = \; a_{ij}^{(r)} \, \kappa^{ij} \tag{4.15}$$

$$a_{ij}^{(r)} \, a_{klm}^{(s)} \, a_n^{(t)} \, \kappa_{\boldsymbol{\xi}} \; = \; a_{ij}^{(r)} \, a_{klm}^{(s)} \, a_n^{(t)} \, \kappa^{ij,klm,n} \; , \; etc... \tag{4.16}$$

where $\kappa^{i_1 \, \cdots \, i_k \, , \, i_{k+1} \, \cdots \, i_{k+l} \, , \, i_{k+l+1} \, \cdots \, i_{k+l+m}}$ is the generalized cummulant of order (k, l, m). Relying on equation (4.12), the generalized cummulants of $\mathbf{u}$ can be obtained. Then, ordinary cummulants and moments may be deduced. Note that symbolic manipulation packages can be quite helpful in carrying out the symbolic algebra involved in these equations.

Alternatively, cross moments between elements of the response vector $\mathbf{u}$ may be obtained by direct manipulations of equation (4.7). For example, the $m + n$ order moment involving $u^{(r)}$ and $u^{(s)}$ is

$$< [u^{(r)}]^m \, [u^{(s)}]^n > \; =$$
$$< \left[\sum_{k=0}^{\infty} a_{\lambda_0 \ldots \lambda_k}^{(r)} \, \Pi_{\lambda_0 \ldots \lambda_k}(\boldsymbol{\xi}) \right]^m \left[\sum_{k=0}^{\infty} a_{\lambda_0 \ldots \lambda_k}^{(s)} \, \Pi_{\lambda_0 \ldots \lambda_k}(\boldsymbol{\xi}) \right]^n > . \tag{4.17}$$

Note that equation (4.17) involves deterministic tensor multiplication and averages of polynomials of orthonormal Gaussian variates, which may be expedited by using the recursion

$$< \xi_{\lambda_1} \ldots \xi_{\lambda_k} > \; = \; \sum_{p=2}^{k} < \xi_{\lambda_1} \, \xi_{\lambda_p} > < \prod_{\substack{m=2 \\ m! = p}} \xi_{\lambda_m} > . \tag{4.18}$$

The expectation of the product on the right hand side of equation (4.18) involves $(k - 2)$ variables. Again, the algebraic manipulations involved may be quite laborious but lend themselves to treatment with a symbolic manipulation program. Specifically, Table (4.1) displays a computer code written for the symbolic manipulation package MACSYMA, which evaluates the averaged products of two-dimensional Polynomial Chaoses up to a certain order. Figure (4.2) shows the code corresponding to the four-dimensional Polynomial Chaoses. It is noted that a trivial modification of these codes is required to permit the evaluation of averaged products of Polynomial Chaoses of arbitrary order. By including additional terms in the expansion (4.7), and random variables in the set $\{\xi_i\}$, any level of accuracy for any order moment can be achieved.

```
\* kronecker : *\

kdelta(x,y)::=buildq([x,y],
if x=y then 1 else 0)$

\* average : *\

average(p)::=
buildq([p],
( logexpand:super,
order:0,
for i: 1 thru n
do ( z:coeff(log(p),log(x[i])),
if z#0 then
for j:1 thru z do y[j+order]:x[i],
order:order+z ),
hom:p/prod(y[i],i,1,order),
if oddp(order) then 0 else
(if order=0 then
hom
else
if order=2 then
hom*kdelta(y[1],y[2])
else
hom*sum(kdelta(y[1],y[i])
   *prod(y[k],k,2,order)/y[i],i,2,order))))$
```

Table 4.1:, MACSYMA Macro to Average Products of
Two-Dimensional Polynomial Chaoses.

```
\* herm2 : *\

ind:0$
G2[0]:1$
gen:exp(-sum(x[i]^2/2,i,1,2))$
for i:1 thru 2 do
(ind:ind+1,
G2[ind]:expand(-diff(gen,x[i])/gen) )$
for i:1 thru 2 do
(for j:i thru 2 do
(ind:ind+1,
G2[ind]:expand(-diff(-diff(gen,x[i]),x[j])/gen)))$
for i:1 thru 2 do
(for j:i thru 2 do
(for k:j thru 2 do
(ind:ind+1,
G2[ind]:expand(-diff(-diff(
        -diff(gen,x[i]),x[j]),x[k])/gen))))$
for i:1 thru 2 do
(for j:i thru 2 do
(for k:j thru 2 do
(for l:k thru 2 do
(ind:ind+1,
G2[ind]:expand(-diff(-diff(-diff(
   -diff(gen,x[i]),x[j]),x[k]),x[l])/gen )))))$
```

Table 4.1, (continued):, MACSYMA Macro to Average Products of
Two-Dimensional Polynomial Chaoses.

```
\* product2 : *\

product2(num)::=(buildq([num],
(
n:2,
load(kronecker),
load(average),
load(herm2),
for i:1 thru num do
(for j:i thru num do
(for k:j thru num do
(pp:expand(G2[i]*G2[j]*G2[k]),
p[0]:pp,
mm:0,
for m:1 while integerp(p[m-1])=false do
(mm:m,
p[m]:if length(p[m-1])=2
and integerp(part(p[m-1],2))=true
and length(part(p[m-1],1))=1
then average(p[m-1])
else map(average,p[m-1])),
c2[i,j,k]:p[mm],
if c2[i,j,k]#0 then print(i,j,k,c2[i,j,k],pp) ))) )))$
```

Table 4.1, (continued):, MACSYMA Macro to Average Products of
Two-Dimensional Polynomial Chaoses.

```
\* kronecker : *\

kdelta(x,y)::=buildq([x,y],
if x=y then 1 else 0)$


\* average : *\

average(p)::=
buildq([p],
( logexpand:super,
order:0,
for i: 1 thru n
do ( z:coeff(log(p),log(x[i])),
if z#0 then
for j:1 thru z do y[j+order]:x[i],
order:order+z ),
hom:p/prod(y[i],i,1,order),
if oddp(order) then 0 else
(if order=0 then
hom
else
if order=2 then
hom*kdelta(y[1],y[2])
else
hom*sum(kdelta(y[1],y[i])
   *prod(y[k],k,2,order)/y[i],i,2,order))))$
```

Table 4.2:, MACSYMA Macro to Average Products of Four-Dimensional Polynomial Chaoses.

```
\* herm4 : *\

ind:0$
G4[0]:1$
gen:exp(-sum(x[i]^2/2,i,1,4))$
for i:1 thru 4 do
(ind:ind+1,
G4[ind]:expand(-diff(gen,x[i])/gen) )$
for i:1 thru 4 do
(for j:i thru 4 do
(ind:ind+1,
G4[ind]:expand(-diff(-diff(gen,x[i]),x[j])/gen)))$
for i:1 thru 4 do
(for j:i thru 4 do
(for k:j thru 4 do
(ind:ind+1,
G4[ind]:expand(-diff(-diff(
        -diff(gen,x[i]),x[j]),x[k])/gen))))$
for i:1 thru 4 do
(for j:i thru 4 do
(for k:j thru 4 do
(for l:k thru 4 do
(ind:ind+1,
G4[ind]:expand(-diff(-diff(-diff(
        -diff(gen,x[i]),x[j]),x[k]),x[l])/gen)))))$
```

Table 4.2, (continued):, MACSYMA Macro to Average Products of
Four-Dimensional Polynomial Chaoses.

```
\* product4 : *\

product4(num)::=(buildq([num],
(
n:4,
load(kronecker),
load(average),
load(herm4),
for i:1 thru num do
(for j:i thru num do
(for k:j thru num do
(pp:expand(G4[i]*G4[j]*G4[k]),
p[0]:pp,
mm:0,
for m:1 while integerp(p[m-1])=false do
(mm:m,
p[m]:if length(p[m-1])=2
and integerp(part(p[m-1],2))=true
and length(part(p[m-1],1))=1
then average(p[m-1])
else map(average,p[m-1])),
c4[i,j,k]:p[mm],
if c4[i,j,k]#0 then print(i,j,k,c4[i,j,k],pp) ))) )))$
```

Table 4.2, (continued):, MACSYMA Macro to Average Products of
Four-Dimensional Polynomial Chaoses.

4.2.2 Second Order Statistics

As it has already been discussed in section (4.2.1), the second order statistics
of response quantities are of particular importance in reliability analysis.
These can provide preliminary estimates of the values of the reliability index
and the probability of failure. The stochastic finite element method as
described in the Chapter III can be used quite efficiently to produce accurate
approximations to these statistics. Specifically, incorporating the improved
Neumann expansion method described in section (3.3.5) in the procedure of
the previous section, one obtains the following equation for the mean of the

response process,

$$\bar{u} \equiv <u> = \left[I + \sum_{i=1}^{M} Q^{(i)} Q^{(i)} \right.$$

$$\left. + \sum_{l=1}^{M} \sum_{n=1}^{M} \sum_{j=1}^{M} \sum_{i=1}^{M} <\xi_i \xi_j \xi_k \xi_l> Q^{(i)} Q^{(j)} Q^{(k)} Q^{(l)} + ... \right] g . \tag{4.19}$$

Further, making use of equation (4.18) in the form

$$<\xi_i \xi_j \xi_k \xi_l> = \delta_{ij} \delta_{kl} + \delta_{ik} \delta_{jl} + \delta_{il} \delta_{jk} , \tag{4.20}$$

equation (4.19) becomes

$$\bar{u} = \left[I + \sum_{i=1}^{M} Q^{(i)} Q^{(i)} \sum_{j=1}^{M} \sum_{i=1}^{M} \left[Q^{(i)} Q^{(i)} Q^{(j)} Q^{(j)} \right. \right.$$

$$\left. \left. + Q^{(i)} Q^{(j)} Q^{(j)} Q^{(i)} + Q^{(i)} Q^{(j)} Q^{(i)} Q^{(j)} \right] + ... \right] g . \tag{4.21}$$

The covariance matrix R_{uu} of the response can then be obtained by evaluating the averaged outer product of the response vector with itself. Specifically,

$$R_{uu} = \sum_{\substack{i=0 \ j=0 \\ i+j even}} < \left[\sum_{m=1}^{M} \xi_m Q^{(m)} \right]^i G \left[\sum_{n=1}^{M} \xi_n Q^{(n)} \right]^j > , \tag{4.22}$$

where

$$G = gg^T . \tag{4.23}$$

Expanding equation (4.22) yields

$$R_{uu} = \left[I + \sum_{i=1}^{M} \sum_{j=1}^{M} \left[Q^{(i)}Q^{(j)}G + Q^{(i)}GQ^{(j)} + GQ^{(i)}Q^{(j)} \right] \right.$$

$$+ \sum_{i=1}^{M} \sum_{j=1}^{M} \sum_{k=1}^{M} \sum_{l=1}^{M} <\xi_i \xi_j \xi_k \xi_l> \left[Q^{(i)} Q^{(j)} Q^{(k)} Q^{(l)} G \right.$$

$$+ Q^{(i)} Q^{(j)} Q^{(k)} G Q^{(l)} + Q^{(i)} Q^{(j)} G Q^{(k)} Q^{(l)}$$

$$+ Q^{(i)} G Q^{(j)} Q^{(k)} Q^{(l)} + G Q^{(i)} Q^{(j)} Q^{(k)} Q^{(l)} \right]$$

$$\left. + ... \right] . \tag{4.24}$$

Equation (4.24) can be simplified by relying on equation (4.18) which, for the the sixth order term becomes,

$$
\begin{aligned}
<\xi_{\lambda_1} \xi_{\lambda_2} \xi_{\lambda_3} \xi_{\lambda_4} \xi_{\lambda_5} \xi_{\lambda_6}> \ = \ & <\xi_{\lambda_1} \xi_{\lambda_2}> < \xi_{\lambda_3} \xi_{\lambda_4} \xi_{\lambda_5} \xi_{\lambda_6}> \\
& + \ <\xi_{\lambda_1} \xi_{\lambda_3}> <\xi_{\lambda_2} \xi_{\lambda_4} \xi_{\lambda_5} \xi_{\lambda_6}> \\
& + \ <\xi_{\lambda_1} \xi_{\lambda_4}> <\xi_{\lambda_2} \xi_{\lambda_3} \xi_{\lambda_5} \xi_{\lambda_6}> \\
& + \ <\xi_{\lambda_1} \xi_{\lambda_5}> <\xi_{\lambda_2} \xi_{\lambda_3} \xi_{\lambda_4} \xi_{\lambda_6}> \\
& + \ <\xi_{\lambda_1} \xi_{\lambda_6}> <\xi_{\lambda_2} \xi_{\lambda_3} \xi_{\lambda_4} \xi_{\lambda_5}> \ .
\end{aligned}
\tag{4.25}
$$

Alternatively, certain properties of the Polynomial Chaoses can be used to simplify considerably the calculation of the associated response statistics. Specifically, recalling that these Polynomials have zero-mean, the mean response vector is given by the equation

$$
\bar{\mathbf{u}} \ \equiv \ <\mathbf{u}> \ = \ \mathbf{d}_0
\tag{4.26}
$$

where the notation conforms to that introduced in section (3.3.6). Further, relying on the orthogonality of the Polynomial Chaoses one can derive the following simple expression for the covariance matrix of the response,

$$
\mathbf{R}_{uu} \ = \ \sum_{j=0}^{P} <\Psi_j[\{\xi_r\}] \ \Psi_j[\{\xi_r\}]> \ \mathbf{d}_j \ \mathbf{d}_j^H \ ,
\tag{4.27}
$$

where the superscript H over a vector denotes its hermitian transpose. In this regard recall that values of the variances of some of the Polynomial Chaoses, appearing in equation (4.27) are displayed in Tables (2.1)-(2.3), whereas MACSYMA symbolic manipulation programs to compute them are shown in Tables (2.5) and (2.6).

4.3 Approximation to the Probability Distribution

The moments computed according to the previous section can be used either directly to estimate probabilities of failure, or in conjunction with some expansion, such as the Edgeworth expansion (McCullagh, 1987), to approximate the probability distribution function of the response process. An alternative approach to obtaining these probability distributions consists of directly estimating the coefficients appearing in their expansion, thus bypassing the moment calculation stage. As a consequence of the general form into which the response process was cast, equation (4.7), it is conceptually possible to obtain approximations to the probability distribution of **u**. Grigoriu (1982) has called attention to methods for estimating the probability distribution of polynomials as weighted averages of some judiciously chosen probability distributions. A different approach is followed herein which is suggested by the fact that the entire probabilistic information concerning the response process is actually condensed in the coefficients of the expansion (4.7). Note that using the Homogeneous Chaos formulation, the polynomials in equation (4.7) are orthogonal. Further, assume without any loss of generality that they are normalized. Then, multiplying equation (4.7) by $\Gamma_{\lambda_0...\lambda_p}(\boldsymbol{\xi})$ and averaging gives

$$<u\ \Gamma_{\lambda_0...\lambda_p}(\boldsymbol{\xi})>\ =\ a_{\lambda_0...\lambda_p}\ , \tag{4.28}$$

where u is used to denote any element of the vector **u**. The expected value in equation (4.28) may be formally written as

$$\int_{-\infty}^{\infty}\ ...\ \int_{-\infty}^{\infty}\ u\ \Gamma_{\lambda_0...\lambda_p}(\boldsymbol{\xi})\ p_{u,\boldsymbol{\xi}}(u,\boldsymbol{\xi})\ du\ d\boldsymbol{\xi}\ =\ a_{\lambda_0...\lambda_p}\ , \tag{4.29}$$

where $p_{u,\boldsymbol{\xi}}(u,\boldsymbol{\xi})$ is an $(K+1)$-dimensional joint distribution density of u and $\boldsymbol{\xi}$. This density function can be expanded in a multidimensional Edgeworth series as

$$p_{u,\boldsymbol{\xi}}(u,\boldsymbol{\xi})\ =\ \sum_{k=0}^{\infty}\ b_{\lambda_0...\lambda_k}\ \Gamma_{\lambda_0...\lambda_k}(u,\boldsymbol{\xi})\ \Phi(u,\boldsymbol{\xi})\ , \tag{4.30}$$

where $\Phi(x,\boldsymbol{\xi})$ is the $(K+1)$-dimensional Gaussian distribution derived from the covariance matrix of the vector $\boldsymbol{\xi}$, augmented by the response u. That is,

$$\Phi(u,\boldsymbol{\xi})\ =\ \frac{1}{(2\pi)^{-(K+1)/2}\ |\mathbf{C}|^{-\frac{1}{2}}}\ e^{-\frac{1}{2}\ (\mathbf{z}-\bar{\mathbf{z}})^T\ \mathbf{C}\ (\mathbf{z}-\bar{\mathbf{z}})}\ , \tag{4.31}$$

where

$$\mathbf{C}\ =\ <\mathbf{z}\mathbf{z}^T>\ ,\ \bar{\mathbf{z}}\ =\ <\mathbf{z}> \tag{4.32}$$

and

$$\mathbf{z}\ =\ \left[\begin{array}{c} u \\ \boldsymbol{\xi} \end{array}\right]\ . \tag{4.33}$$

Clearly ξ_i represents a first order Polynomial Chaos, it is orthogonal to all the other Polynomial Chaoses. Consequently, the covariance matrix in equation (4.32) can be written as

$$\mathbf{C} \;=\; \begin{bmatrix} \sigma_u^2 & a_1 & \cdots & a_k \\ a_1 & & & \\ \vdots & & \mathbf{I_K} & \\ a_k & & & \end{bmatrix} \tag{4.34}$$

where $\mathbf{I}_K$ is the Kth order identity matrix, σ_u is the standard deviation of the response process, and a_i is the coefficient of ξ_i in equation (4.7). Substituting equations (4.7) and (4.30) into equation (4.29) gives

$$\sum_{k=0}^{\infty} a_{\lambda_0\ldots\lambda_k} \sum_{p=0}^{\infty} b_{\lambda_0\ldots\lambda_p} \tag{4.35}$$

$$\int_{-\infty}^{\infty} \Gamma_{\lambda_0\ldots\lambda_k}(\boldsymbol{\xi})\,\Gamma_{\lambda_0\ldots\lambda_m}(\boldsymbol{\xi})\,\Gamma_{\lambda_0\ldots\lambda_p}(u,\boldsymbol{\xi})\,\Phi(u,\boldsymbol{\xi})\,du\,d\boldsymbol{\xi} \;=\; a_{\lambda_0\ldots\lambda_k}\,.$$

Note that the integrands in equation (4.35) involve products of multidimensional polynomials with the Gaussian distribution. Thus, the value of the integral is equal to sums of moments of the multidimensional normal distribution given by equation (4.31). This distribution is completely determined by the joint second order moments of the response process and the variables $\{\xi_i\}$. These moments can be readily obtained by multiplying equation (4.7) by ξ_i upon averaging. Equation (4.35) can be cast in a matrix form

$$\mathbf{A}\,\mathbf{b} \;=\; \mathbf{a}\,, \tag{4.36}$$

where $\mathbf{b}$ is the vector of unknown coefficients in equation (4.30), $\mathbf{a}$ is the vector of known coefficients in equation (4.7), and $\mathbf{A}$ is a rectangular matrix. Note that the polynomials used in equation (4.30) have $K+1$ variables to accommodate the response u, whereas the polynomials used in the expansion (4.30) have only K variables. Therefore the number of unknowns is greater than the number of available equations. In this regard, additional equations may be obtained by requiring the vector $\boldsymbol{\xi}$ to be jointly normal. That is,

$$\int_{-\infty}^{\infty} p_{u,\boldsymbol{\xi}}(u,\boldsymbol{\xi})\,du \;=\; \Phi(\boldsymbol{\xi})\,, \tag{4.37}$$

where $\Phi(\boldsymbol{\xi})$ is the K-dimensional normal distribution of independent identically distributed, zero mean Gaussian variables. Using these additional
conditions, equation (4.36) can be augmented to involve a coefficient matrix
which is square. Solving the augmented matrix equation, the vector $\mathbf{b}$ can be
found and used in equation (4.30) to express $p_{u,\boldsymbol{\xi}}(u, \boldsymbol{\xi})$. Finally, note that
upon determining the vector $\mathbf{b}$, the marginal distribution of the response
process may also be obtained from equation (4.30) by integration over the
$\boldsymbol{\xi}$-space. That is,

$$p_u(u) = \int_{-\infty}^{\infty} p_{u,\boldsymbol{\xi}}(u, \boldsymbol{\xi}) \, d\boldsymbol{\xi} \,. \tag{4.38}$$

4.4 Reliability Index and Response Surface Simulation

As defined in the introductory section to this Chapter, the generalized
reliability index is the shortest distance to the limit state surface from the
mean design variables. The point on the limit surface corresponding to this
shortest distance was described as being the most probable failure point.
Equation (4.7) with the left hand side replaced by $u_{failure}$ may be regarded
as an explicit expression of the failure surface in the space spanned by the ξ_i.
This surface is obviously nonlinear. Standard optimization algorithms may
be used to compute the reliability index and the corresponding probable-
failure point. The Karhunen-Loeve expansion may then be used to recover
the physical value of the failure point in the space spanned by the random
process itself.

Computing the reliability index, however, is unnecessary, as the calculation of the probability distribution of the response is itself within reach.
The latter is best evaluated by simulating points on the response surface.
Clearly, once the coefficients for the representation of the response process
in the form (4.7) have been computed, it becomes possible to obtain points
on the response surface using simulation techniques. In fact, to each real-
ization of the set $\{\xi_{\lambda_0}, ..., \xi_{\lambda_p}\}$ there corresponds a unique realization of the
response process $u(\mathbf{x}, \theta)$. Clearly, this procedure yields a sample population
for the response process which can be used to estimate its probability density
function. In this context, Non-parametric kernel density estimation methods
(Beckers and Chambers, 1984) are particularly useful.

Chapter 5

NUMERICAL EXAMPLES

5.1 Preliminary Remarks

The methods introduced in sections (3.3.5) and (3.3.6) are exemplified in this chapter by considering three problems reflecting some interesting applications from engineering mechanics. These problems involve a beam with random rigidity, a plate with random rigidity, and a beam resting on a random elastic foundation and subjected to a random dynamic excitation. In addressing these problems, it is reminded that the ultimate goal of a stochastic finite element analysis is the calculation of certain statistics of the response process. These statistics can be in the form of either statistical moments, or probability distribution function, or some other measure of the reliability of the system. As a first step in the solution procedure, the variational formulation of the finite element method is used to obtain a spatially discrete form of the problem. Following that, the Neumann expansion for the inverse, as given by equation (3.46), and the Polynomial Chaos expansion, as given by equation (3.81), are used to derive a representation of the response process. Statistical moments and probability distribution functions are then obtained as discussed in Chapter IV.

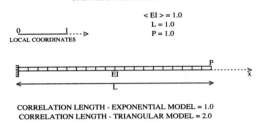

Figure 5.1: Beam with Random Bending Rigidity Under Uniform Load; Exponential and Triangular Covariance Models.

5.2 One Dimensional Static Problem

5.2.1 Formulation

Consider the Euler Bernoulli beam shown in Figure (5.1), of length L, clamped at one end and subjected to a deterministic static transverse load P. It is assumed that the bending rigidity $w \equiv EI$ of the beam, which involves the modulus of elasticity E and the cross-sectional mass moment of inertia I, is the realization of a Gaussian random process indexed over the spatial domain occupied by the beam. It is also assumed that the random bending rigidity w is represented by its mean value $\bar{w}$ and its covariance function $C(\mathbf{x}_1, \mathbf{x}_2)$.

The strain energy V stored in the beam can be written as

$$V = \frac{1}{2} \int_L \boldsymbol{\sigma}(x) \, \boldsymbol{\epsilon}(x) \, dx , \qquad (5.1)$$

where $\boldsymbol{\sigma}(x)$ and $\boldsymbol{\epsilon}(x)$ represent the stress and the strain as a function of the location on the beam. Assuming the material to be linear, and making use of Hooke's law, equation (5.1) becomes

$$V = \frac{1}{2} \int_L w(x,\theta) \left(\frac{\partial^2}{\partial x^2} u(x,\theta) \right)^2 dx, \qquad (5.2)$$

where $w(x,\theta)$ is the random bending rigidity as described above, and $u(x,\theta)$ is the random transverse displacement of the beam. It is reminded here that $\theta \in \Omega$, where Ω denotes the space of random events. At this point, $u(x,\theta)$ may be expanded, as a function of x, along a basis of piecewise polynomials as follows

$$u(x,\theta) = \sum_{e=1}^{N} \mathbf{u}_e(\theta) \, \mathbf{p}_e^{(n_e)}(x) \,, \qquad (5.3)$$

where N is the total number of degrees of freedom in the discrete model. In this equation, $\mathbf{p}_e^{(n_e)}(x)$ is a vector of polynomials that are identically zero except over a subdomain l^e where they are of order n_e such that certain compatibility conditions are satisfied on the boundaries of l^e. Further, $\mathbf{u}_e(\theta)$ is a vector of unknown random coefficients, as indicated by the argument θ, representing the transverse displacement and slope at some point in l^e. The subscript e on the superscript n provides for different order polynomials over different subdomains. These subdomains represent the induced finite element mesh over the beam. Note that the order n_e of the polynomials used in this problem needs to be at least equal to two for the integrand in equation (5.2) to remain finite. Substituting equation (5.3) into equation (5.2) gives

$$V = \frac{1}{2} \sum_{e=1}^{N} \sum_{s=1}^{N} \int_L \left(\mathbf{u}_e(\theta) \frac{d^2}{dx^2} \mathbf{p}_e^{(n_e)}(x) \right) w(x,\theta) \qquad (5.4)$$

$$\left(\mathbf{u}_s(\theta) \frac{d^2}{dx^2} \mathbf{p}_s^{(n_f)}(x) \right) dx \,.$$

In a similar manner, the work performed by the applied force may be put in the form

$$V' = \sum_{e=1}^{N} \mathbf{u}_e(\theta) \int_L \mathbf{p}_e^{(n_e)}(x) \, f(x) \, dx \,. \qquad (5.5)$$

The Minimization of the total potential energy ($V - V'$) can be expressed mathematically as

$$\frac{\partial\,(\,V\,-\,V'\,)}{\partial\mathbf{u}_e(\theta)} \;=\; 0 \;. \tag{5.6}$$

Substituting equations (5.4) and (5.5) into equation (5.6) gives

$$\sum_{e=1}^{N} \mathbf{u}_e(\theta) \int_L \left(\frac{d^2}{dx^2}\mathbf{p}_e^{(n_e)}(x) \right) w(x,\theta) \left(\frac{d^2}{dx^2}\mathbf{p}_s^{(n_f)}(x) \right) dx$$

$$= \int_L \mathbf{p}_s^{(n_f)}(x)\, f(x)\, dx \;,\; s = 1,...,N \;. \tag{5.7}$$

Evaluating equation (5.7) for all $s \in [1, N]$, leads to

$$\mathbf{K}\,\mathbf{u} \;=\; \mathbf{f} \;, \tag{5.8}$$

where

$$\mathbf{K}_{es} \;=\; \int_L \left(\frac{d^2}{dx^2}\mathbf{p}_e^{(n_e)}(x) \right) w(x,\theta) \left(\frac{d^2}{dx^2}\mathbf{p}_s^{(n_f)}(x) \right) dx \tag{5.9}$$

$$\mathbf{u}_e \;=\; \mathbf{u}_e(\theta) \tag{5.10}$$

and

$$\mathbf{f}_e \;=\; \int_L \mathbf{p}_e^{(n_e)}(x)\, f(x)\, dx \;. \tag{5.11}$$

Note that, since the polynomials $\mathbf{p}_e^{(n_e)}(x)$ have bounded support, the integrals over L in the above equations may be replaced by integrals over the corresponding supporting subdomains. Further, these integrals involve the random process $w(x,\theta)$, whose spatial variation is not suitably defined for the purpose of carrying out the integrations. This problem can be resolved by expanding $w(x,\theta)$ in its truncated Karhunen-Loeve series. Specifically,

$$w(x,\theta) \;=\; \bar{w}(x) \;+\; \sum_{k=1}^{M} \xi_k(\theta)\,\sqrt{\lambda_k}\,f_k(x) \;, \tag{5.12}$$

where $\xi_k(\theta)$, λ_k and $f_k(x)$ are as defined in section (2.3). Substituting for $w(x,\theta)$, equation (5.8) becomes

$$\sum_{k=0}^{M} \xi_k(\theta)\,\mathbf{K}^{(k)}\,\mathbf{u} \;=\; \mathbf{f} \;, \tag{5.13}$$

where

$$\mathbf{K}_{es}^{(k)} = \int_L \sqrt{\lambda_k} \, f_k(x) \left(\frac{d^2}{dx^2} \mathbf{p}_e^{(n_e)}(x) \right) \left(\frac{d^2}{dx^2} \mathbf{p}_s^{(n_f)}(x) \right)$$
$$dx \, , \ k = 1, ..., M, \tag{5.14}$$

$$\mathbf{K}_{es}^{(0)} = \int_L \bar{w}(x) \left(\frac{d^2}{dx^2} \mathbf{p}_e^{(n_e)}(x) \right) \left(\frac{d^2}{dx^2} \mathbf{p}_s^{(n_f)}(x) \right) \, dx \, , \tag{5.15}$$

and

$$\xi_0 \equiv 1 \, . \tag{5.16}$$

In the remainder of this chapter, the argument θ will be dropped except where it is needed to emphasize the random nature of a certain quantity. The integration in equation (5.14) may be performed either analytically if the eigenfunctions of the covariance kernel are known, or using a quadrature scheme if the eigenvectors have been computed numerically at discrete spatial points. To expedite the calculation of the above integrals, a change of variable is introduced involving the local coordinate

$$r = \frac{(x - x_l)}{l^e} \, , \tag{5.17}$$

where x_l denotes the coordinate of the left end of the element (e). Then, equation (5.14) can be viewed as a combination of integrals of the form

$$I_1^{(k)} = \int_0^1 f_k(rl^e + x_l) \, r^n \, dr \, , \tag{5.18}$$

where n is some integer. Further, the covariance kernel of the random process is modeled using both the exponential model defined by equation (2.39) and the triangular model defined by equation (2.56). For the exponential model, after substituting for $f_k(x)$ from equation (2.51) and (2.52), equation (5.18) becomes

$$
I_1^{(k)} = \begin{bmatrix} \displaystyle\int_0^1 \frac{\cos(\omega_k \ (rl^e \ + \ x_l))}{\sqrt{a + \dfrac{\sin(2\omega_k a)}{2\omega_k}}} \ r^n \ dr \ , \quad k \ even \\[30pt] \displaystyle\int_0^1 \frac{\sin(\omega_k \ (rl^e \ + \ x_l))}{\sqrt{a - \dfrac{\sin(2\omega_k a)}{2\omega_k}}} \ r^n \ dr \ , \quad k \ odd \end{bmatrix} \tag{5.19}
$$

where $a = L/2$ and $c = 1/b$, and b is the correlation length.

For the triangular kernel, the eigenfunctions are given by equations (2.62) and (2.63), and equation (5.18) becomes

$$
I_1^{(k)} = \begin{bmatrix} \displaystyle\int_0^1 \frac{\cos(\omega_k(rl^e + x_l)) \ + \ \tan(\frac{\omega_k \ a}{2}) \ \sin(\omega_k(rl^e + x_l))}{\sqrt{a + \dfrac{\tan^2(\frac{\omega_k \ a}{2}) \ (a - \sin(2\omega_k a))}{4\omega_k} + \dfrac{\sin^2(\omega_k a)}{\omega_k} \ \tan(\frac{\omega_k a}{2})}} \ r^n \ dr \\[30pt] \qquad\qquad\qquad\qquad\qquad k \ even \\[20pt] \displaystyle\int_0^1 \frac{\cos(\omega_k(rl^e + x_l))}{\sqrt{\dfrac{a}{2} + \dfrac{\sin(2\omega_k a)}{2\omega_k}}} \ r^n \ dr \ , \qquad k \ odd \end{bmatrix} \tag{5.20}
$$

Relying on equations (5.19) or (5.20), closed form expressions for the elements of the matrix $\mathbf{K}$ defined by equation (5.14) are derived. To simplify the algebraic manipulations, equation (5.13) is rewritten as

$$
\left[\mathbf{I} + \sum_{k=1}^{M} \xi_k(\theta) \ \mathbf{Q}^{(k)} \right] \mathbf{u} \ = \ \mathbf{g} \ , \tag{5.21}
$$

where

$$
\mathbf{Q}^{(k)} \ = \ [\mathbf{K}^{(0)}]^{-1} \ \mathbf{K}^{(k)} \ , \tag{5.22}
$$

and

$$
\mathbf{g} \ = \ [\mathbf{K}^{(0)}]^{-1} \ \mathbf{f} \ . \tag{5.23}
$$

Due to the boundary conditions imposed at the clamped end of the beam, the first two elements of the vector $\mathbf{u}$ are equal to zero. These elements represent the displacement and the slope at the fixed end, respectively. Substructuring the matrices in equation (5.21) to reflect the boundary conditions leads to

$$\begin{bmatrix} \mathbf{I}_2 \\ & \mathbf{I}_{N-2} \end{bmatrix} + \sum_{k=1}^{M} \xi_k \begin{bmatrix} \mathbf{Q}_{11}^{(k)} & \mathbf{Q}_{12}^{(k)} \\ \mathbf{Q}_{21}^{(k)} & \mathbf{Q}_{22}^{(k)} \end{bmatrix} \begin{bmatrix} \mathbf{u}_1 \equiv 0 \\ \mathbf{u}_2 \end{bmatrix} = \begin{bmatrix} \mathbf{g}_1 \\ \mathbf{g}_2 \end{bmatrix} , \quad (5.24)$$

The substructuring is performed such that $\mathbf{u}_2$ represents the displacements and the rotations of the beam at the free nodes. The 2×1 vector $\mathbf{g}_1$ corresponds to the reactions at the fixed end and may be recovered once the vector $\mathbf{u}_2$ is computed. Expanding equation (5.24) gives

$$\left[\mathbf{I} + \sum_{k=1}^{M} \xi_k \, \mathbf{Q}_{22}^{(k)} \right] \mathbf{u}_2 = \mathbf{g}_2 \quad (5.25)$$

and

$$\left[\mathbf{I} + \sum_{k=1}^{M} \xi_k \, \mathbf{Q}_{12}^{(k)} \right] \mathbf{u}_2 = \mathbf{g}_1 \quad (5.26)$$

Upon solving equation (5.25) for $\mathbf{u}_2$, equation (5.26) yields a direct expression for the unknown reactions.

In the numerical implementation of the preceding analysis which produced the following numerical results, piecewise cubic polynomials were used. That is, $n_e = 3$. Further, the beam was discretized into ten finite elements, resulting in twenty two $(N = 22)$ degrees of freedom.

5.2.2 Results

Improved Neumann Expansion

Following the discussion in section (3.3.5), the response $\mathbf{u}_2$ may be expressed as

$$\mathbf{u}_2 = \sum_{n=0}^{\infty} (-1)^n \left[\sum_{k=1}^{M} \xi_k \, \mathbf{Q}_{22}^{(k)} \right]^n \mathbf{g}_2 . \quad (5.27)$$

The average response may be expressed as in equation (4.19) and the covariance matrix of the response as in equation (4.22). Figure (5.2) shows the

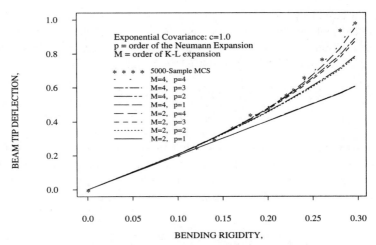

Figure 5.2: Beam Tip Deflection Normalized Standard Deviation versus Bending Rigidity Standard Deviation; Exponential Covariance; $\sigma_{T,max} = 0.297$; Neumann Expansion Solution.

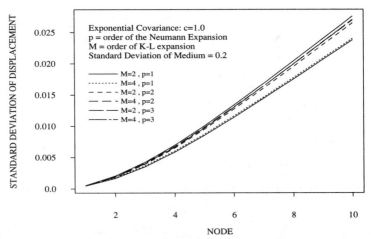

Figure 5.3: Standard Deviation of the Displacement Along the Beam; Numerical Values Determined Only for the Ten Nodes; Exponential Covariance; Neumann Expansion Solution; $\sigma_{EI} = 0.2$.

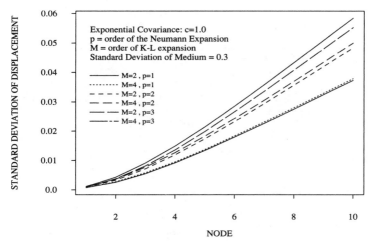

Figure 5.4: Standard Deviation of the Displacement Along the Beam; Numerical Values Determined Only for the Ten Nodes; Exponential Covariance; Neumann Expansion Solution; $\sigma_{EI} = 0.3$.

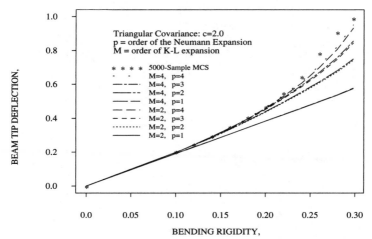

Figure 5.5: Beam Tip Deflection Normalized Standard Deviation versus Bending Rigidity Standard Deviation; Triangular Covariance; $\sigma_{T,max} = 0.298$; Neumann Expansion Solution.

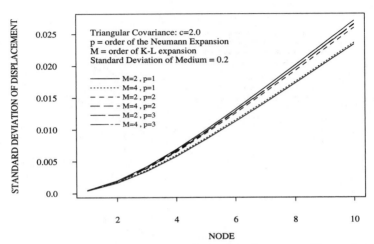

Figure 5.6: Standard Deviation of the Displacement Along the Beam; Numerical Values Determined Only for the Ten Nodes; Triangular Covariance; Neumann Expansion Solution; $\sigma_{EI} = 0.2$.

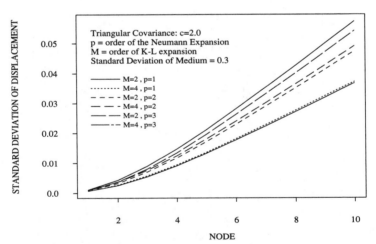

Figure 5.7: Standard Deviation of the Displacement Along the Beam; Numerical Values Determined Only for the Ten Nodes; Triangular Covariance; Neumann Expansion Solution; $\sigma_{EI} = 0.3$.

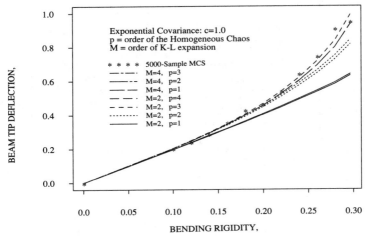

Figure 5.8: Beam Tip Deflection Normalized Standard Deviation versus Bending Rigidity Standard Deviation; Exponential Covariance; $\sigma_{T,max} = 0.297$; Polynomial Chaos Solution.

standard deviation of the displacement at the tip of the beam, σ_u, plotted against the standard deviation of bending rigidity of the beam, σ_{EI}, for several values of the order p of the Neumann expansion and for various values of the order M of the K-L expansion. Figures (5.3)-(5.4) show the standard deviation of the response along the beam, again for various values of p and M. These figures show the results corresponding to the exponentially decaying covariance model. Figures (5.5)-(5.7) show the corresponding results for the triangular model.

Projection on the Homogeneous Chaos

As discussed in section (3.3.6), the response process may also be expanded as in equation (3.81). Incorporating this expansion into equation (5.26) gives

$$\sum_{i=0}^{P} \left[\mathbf{I} + \sum_{k=1}^{M} \xi_k \, \mathbf{Q}_{22}^{(k)} \right] \mathbf{c}_i \, \Psi_i[\{\xi_l\}] = \mathbf{g}_2 \, . \qquad (5.28)$$

Multiplying equation (5.28) by $\Psi_j[\{\xi_l\}]$ and taking the mathematical expec-

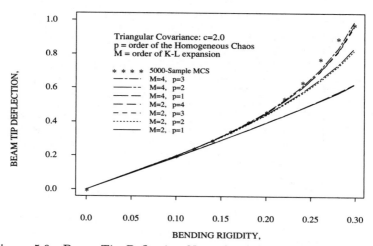

Figure 5.9: Beam Tip Deflection Normalized Standard Deviation versus Bending Rigidity Standard Deviation; Triangular Covariance; $\sigma_{T,max} = 0.298$; Polynomial Chaos Solution.

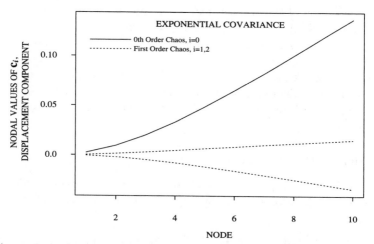

Figure 5.10: Linear Interpolation of the Nodal Values of the Vector $\mathbf{c}_i$ of Equation (5.29) for the Beam Bending Problem, $i = 0, 1, 2$; Displacement Representation; 2 Terms in K-L Expansion, $M = 2$; $p = 0, 1$.

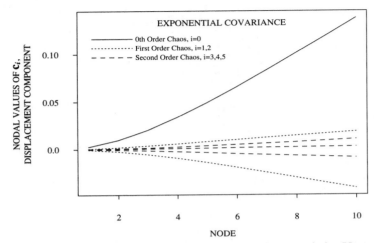

Figure 5.11: Linear Interpolation of the Nodal Values of the Vector $\mathbf{c}_i$ of Equation (5.29) for the Beam Bending Problem, $i = 0, ..., 5$; Displacement Representation; 2 Terms in K-L Expansion, $M = 2$; $p = 0, 1, 2$.

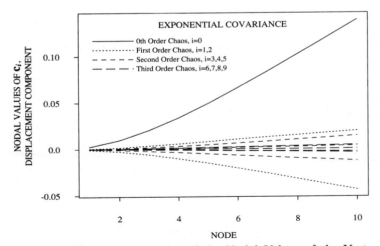

Figure 5.12: Linear Interpolation of the Nodal Values of the Vector $\mathbf{c}_i$ of Equation (5.29) for the Beam Bending Problem, $i = 0, ..., 9$; Displacement Representation; 2 Terms in K-L Expansion, $M = 2$; $p = 0, 1, 2, 3$.

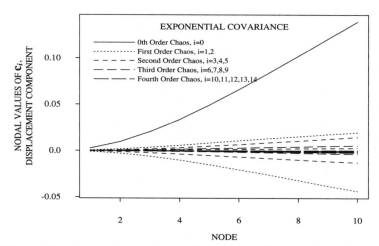

Figure 5.13: Linear Interpolation of the Nodal Values of the Vector $\mathbf{c}_i$ of Equation (5.29) for the Beam Bending Problem, $i = 0, ..., 14$; Displacement Representation; 2 Terms in K-L Expansion, $M = 2$; $p = 0, 1, 2, 3, 4$.

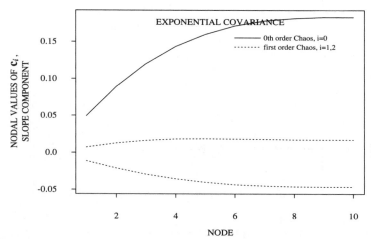

Figure 5.14: Linear Interpolation of the Nodal Values of the Vector $\mathbf{c}_i$ of Equation (5.29) for the Beam Bending Problem, $i = 0, 1, 2$; Slope Representation; 2 Terms in K-L Expansion, $M = 2$; $p = 0, 1$.

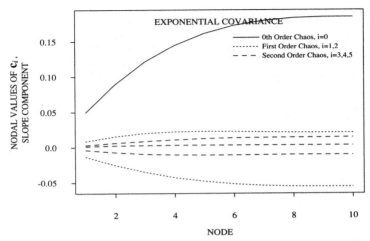

Figure 5.15: Linear Interpolation of the Nodal Values of the Vector $\mathbf{c}_i$ of Equation (5.29) for the Beam Bending Problem, $i = 0, ..., 5$; Slope Representation; 2 Terms in K-L Expansion, $M = 2$; $p = 0, 1, 2$.

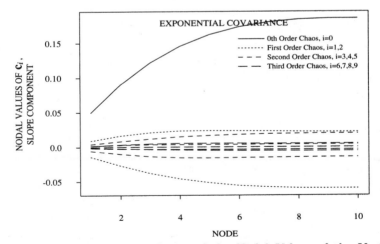

Figure 5.16: Linear Interpolation of the Nodal Values of the Vector $\mathbf{c}_i$ of Equation (5.29) for the Beam Bending Problem, $i = 0, ..., 9$; Slope Representation; 2 Terms in K-L Expansion, $M = 2$; $p = 0, 1, 2, 3$.

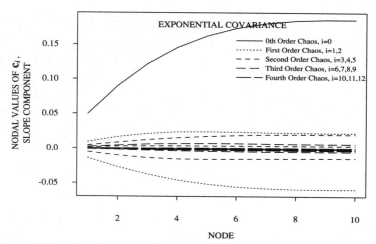

Figure 5.17: Linear Interpolation of the Nodal Values of the Vector c_i of Equation (5.29) for the Beam Bending Problem, $i = 0, ..., 14$; Slope Representation; 2 Terms in K-L Expansion, $M = 2$; $p = 0, 1, 2, 3, 4$.

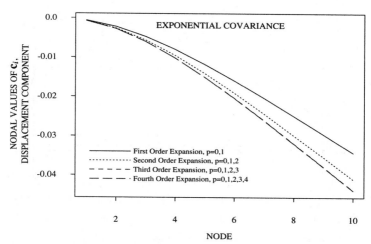

Figure 5.18: Linear Interpolation of the Nodal Values of the Vector c_i of Equation (5.29) for the Beam Bending Problem, $i = 1$; Displacement Representation; 2 Terms in K-L Expansion, $M = 2$; $P = 3, 6, 10, 15$.

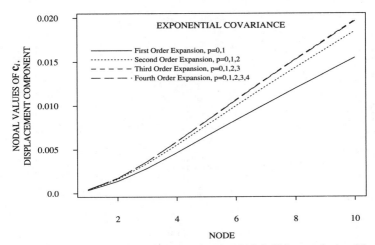

Figure 5.19: Linear Interpolation of the Nodal Values of the Vector $\mathbf{c}_i$ of Equation (5.29) for the Beam Bending Problem, $i = 2$; Displacement Representation; 2 Terms in K-L Expansion, $M = 2$; $P = 3, 6, 10, 15$.

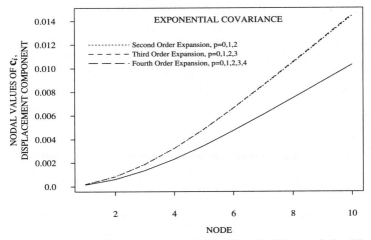

Figure 5.20: Linear Interpolation of the Nodal Values of the Vector $\mathbf{c}_i$ of Equation (5.29) for the Beam Bending Problem, $i = 3$; Displacement Representation; 2 Terms in K-L Expansion, $M = 2$; $P = 6, 10, 15$.

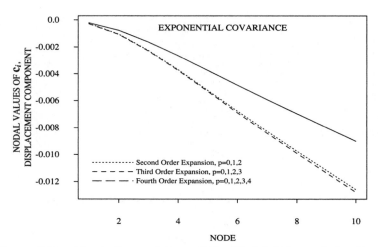

Figure 5.21: Linear Interpolation of the Nodal Values of the Vector $\mathbf{c}_i$ of Equation (5.29) for the Beam Bending Problem, $i = 4$; Displacement Representation; 2 Terms in K-L Expansion, $M = 2$; $P = 6, 10, 15$.

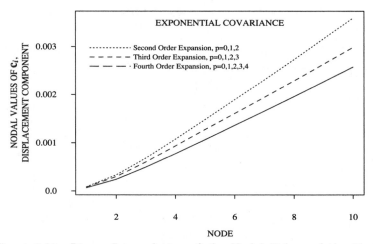

Figure 5.22: Linear Interpolation of the Nodal Values of the Vector $\mathbf{c}_i$ of Equation (5.29) for the Beam Bending Problem, $i = 5$; Displacement Representation; 2 Terms in K-L Expansion, $M = 2$; $P = 6, 10, 15$.

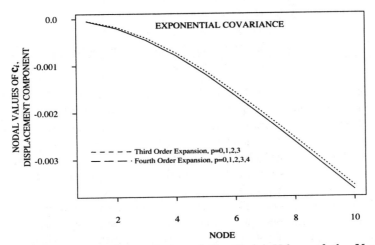

Figure 5.23: Linear Interpolation of the Nodal Values of the Vector $\mathbf{c}_i$ of Equation (5.29) for the Beam Bending Problem, $i = 6$; Displacement Representation; 2 Terms in K-L Expansion, $M = 2$; $P = 10, 15$.

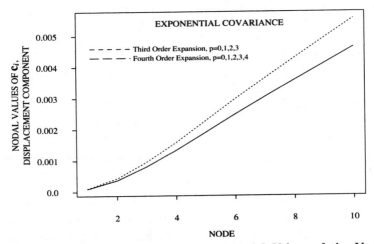

Figure 5.24: Linear Interpolation of the Nodal Values of the Vector $\mathbf{c}_i$ of Equation (5.29) for the Beam Bending Problem, $i = 7$; Displacement Representation; 2 Terms in K-L Expansion, $M = 2$; $P = 10, 15$.

tation, or equivalently requiring the representation error to be orthogonal to the space spanned by the approximating Polynomial Chaoses, gives

$$\left[\sum_{k=0}^{M} \sum_{i=0}^{P} <\xi_k\ \Psi_i[\{\xi_l\}]\ \Psi_j[\{\xi_l\}]> \mathbf{Q}_{22}^{(k)} \right]\ \mathbf{c}_i\ =\ <\Psi_j[\{\xi_l\}]> \mathbf{g}_2\ , \quad (5.29)$$

where $\xi_0 = 1$ and $\mathbf{Q}_{22}^{(0)} = \mathbf{I}$. Recalling the discussion in section (3.3.6) it becomes clear that equation (5.29) leads to a set of algebraic equations to be solved for the vectors $\mathbf{c}_i$. Figures (5.8) and (5.9) show the standard deviation of the displacement at the tip of the beam versus the standard deviation of the bending rigidity. Comparing figures (5.2) and (5.5) to figures (5.8) and (5.9), it is observed that a third order approximation with the Polynomial Chaoses achieves a convergence comparable to that of a fourth order Neumann expansion. The vectors $\mathbf{c}_i$ represent the magnitude of the projections of the response process $u(x, \theta)$ onto the spaces spanned by the successive Polynomial Chaoses $\Psi_i[\{\xi_l\}]$. Figures (5.10)-(5.13) show these projections for the displacement process, corresponding to the exponential covariance model, for a value of the coefficient of variation of the bending rigidity equal to 0.3 and for a correlation length equal to 1. Note the rapid decrease in the magnitude of the projection on the higher order polynomials. Figures (5.14)-(5.17) show similar results pertaining to the slope process along the beam. Due to the term $<\xi_k\ \Psi_i[\{\xi_r\}]\ \Psi_j[\{\xi_r\}]>$ in equation (5.29), the resulting system of equations, although sparse, is coupled. This fact causes the projections to require updating as more terms are included in the summations of equation (5.29). The convergence of these projections with increasing order of the Polynomial Chaos used is shown in Figures (5.18)-(5.26) for the displacement process and in Figures (5.27)-(5.35) for the slope process. Observe the excellent convergence achieved by the leading projections with only a second order expansion. Note that these leading terms provide the largest contribution to the response process. The figures suggest that a reasonable approximation may be obtained by assuming, for higher order polynomials, equal contributions from polynomials of the same order. This assumption may then be relaxed sequentially as more accuracy is required. Once the coefficients in the Polynomial Chaos expansion have been computed, a statistical population corresponding to the response process can be readily generated upon noting that to each realization of the set of random variables $\{\xi_i\}$, there corresponds a realization of the response vector. Realizations of these random variables are obtained from realizations

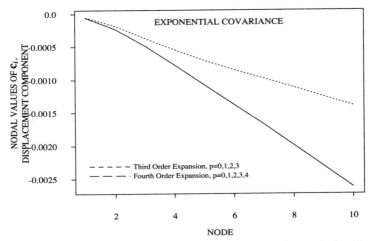

Figure 5.25: Linear Interpolation of the Nodal Values of the Vector $\mathbf{c}_i$ of Equation (5.29) for the Beam Bending Problem, $i = 8$; Displacement Representation; 2 Terms in K-L Expansion, $M = 2$; $P = 10, 15$.

of the random process representing the material variability using equation (2.20). For Gaussian processes, these variables are uncorrelated zero-mean Gaussian random variables with unit variance. Using non-parametric density estimation techniques, as noted in section (4.4), the probability distribution function of the response can be approximated. Figures (5.36)-(5.43) show the probability distributions, obtained as described above, using two, and four terms in the Karhunen-Loeve expansion for the variability of the system. Up to third order Polynomial Chaos is used in the associated computations.

Monte Carlo Simulation

To assess the validity of the results obtained from the analytical methods, and to test their convergence properties, the same problem is treated by the Monte Carlo simulation method. Realizations of the bending rigidity of the beam are numerically simulated using digital filtering techniques (Spanos and Hansen, 1981; Spanos and Mignolet, 1986). An autoregressive filter of order twenty is designed to match the spectral density of the process. As expected, only the first coefficient of the filter is non-negligible, effectively

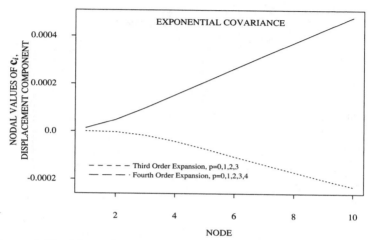

Figure 5.26:　Linear Interpolation of the Nodal Values of the Vector $\mathbf{c}_i$ of Equation (5.29) for the Beam Bending Problem, $i = 9$; Displacement Representation; 2 Terms in K-L Expansion, $M = 2$; $P = 10, 15$.

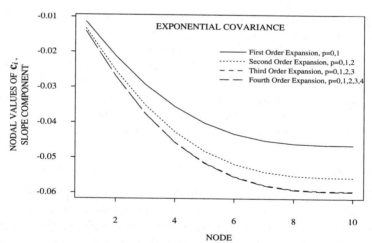

Figure 5.27:　Linear Interpolation of the Nodal Values of the Vector $\mathbf{c}_i$ of Equation (5.29) for the Beam Bending Problem, $i = 1$; Slope Representation; 2 Terms in K-L Expansion, $M = 2$; $P = 3, 6, 10, 15$.

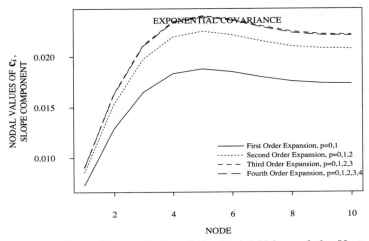

Figure 5.28: Linear Interpolation of the Nodal Values of the Vector $\mathbf{c}_i$ of Equation (5.29) for the Beam Bending Problem, $i = 2$; Slope Representation; 2 Terms in K-L Expansion, $M = 2$; $P = 3, 6, 10, 15$.

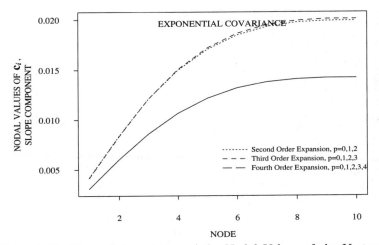

Figure 5.29: Linear Interpolation of the Nodal Values of the Vector $\mathbf{c}_i$ of Equation (5.29) for the Beam Bending Problem, $i = 3$; Slope Representation; 2 Terms in K-L Expansion, $M = 2$; $P = 6, 10, 15$.

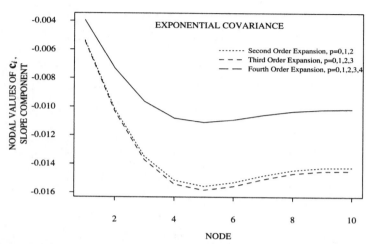

Figure 5.30: Linear Interpolation of the Nodal Values of the Vector c_i of Equation (5.29) for the Beam Bending Problem, $i = 4$; Slope Representation; 2 Terms in K-L Expansion, $M = 2$; $P = 6, 10, 15$.

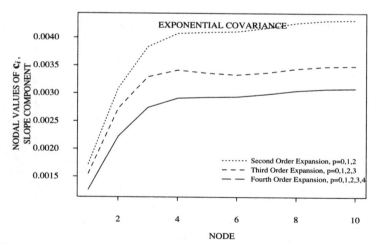

Figure 5.31: Linear Interpolation of the Nodal Values of the Vector c_i of Equation (5.29) for the Beam Bending Problem, $i = 5$; Slope Representation; 2 Terms in K-L Expansion, $M = 2$; $P = 6, 10, 15$.

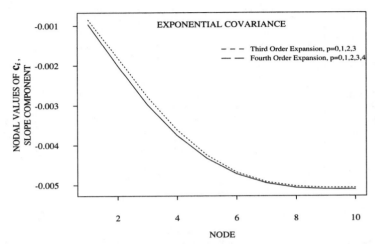

Figure 5.32: Linear Interpolation of the Nodal Values of the Vector $\mathbf{c}_i$ of Equation (5.29) for the Beam Bending Problem, $i = 6$; Slope Representation; 2 Terms in K-L Expansion, $M = 2$; $P = 10, 15$.

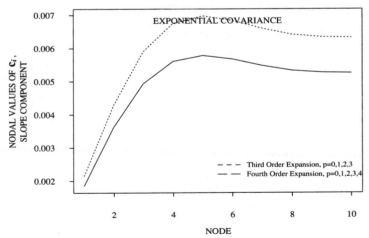

Figure 5.33: Linear Interpolation of the Nodal Values of the Vector $\mathbf{c}_i$ of Equation (5.29) for the Beam Bending Problem, $i = 7$; Slope Representation; 2 Terms in K-L Expansion, $M = 2$; $P = 10, 15$.

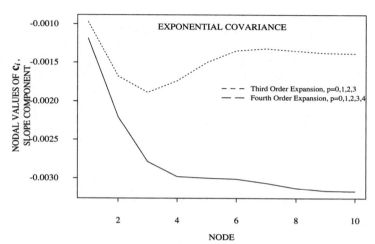

Figure 5.34: Linear Interpolation of the Nodal Values of the Vector c_i of Equation (5.29) for the Beam Bending Problem, $i = 8$; Slope Representation; 2 Terms in K-L Expansion, $M = 2$; $P = 10, 15$.

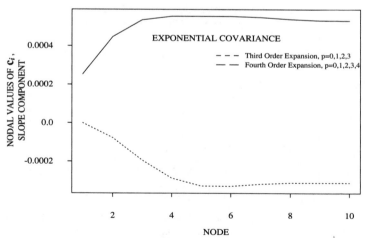

Figure 5.35: Linear Interpolation of the Nodal Values of the Vector c_i of Equation (5.29) for the Beam Bending Problem, $i = 9$; Slope Representation; 2 Terms in K-L Expansion, $M = 2$; $P = 10, 15$.

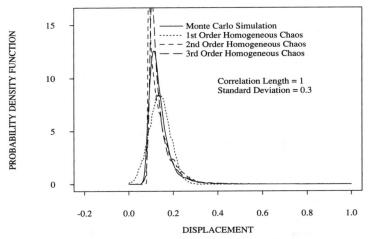

Figure 5.36: Probability Density Function of the Displacement at the Tip of the Beam Using 10,000-Sample Monte Carlo Simulation and up to Third Order Homogeneous Chaos; Two Terms in the K-L Expansion.

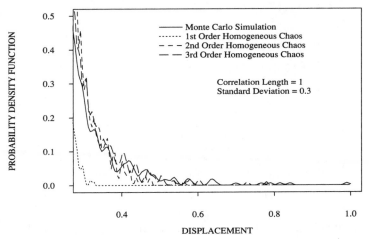

Figure 5.37: Tail of the Probability Density Function of the Displacement at the Tip of the Beam Using 10,000-Sample Monte Carlo Simulation and up to Third Order Homogeneous Chaos; Two Terms in the K-L Expansion.

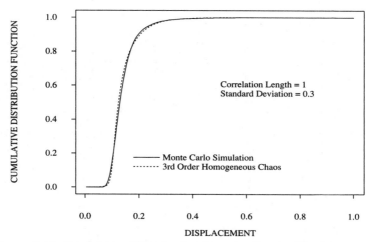

Figure 5.38: Cumulative Distribution Function for the Displacement at the Tip of the Beam Using 10,000-Sample Monte Carlo Simulation and a Third Order Homogeneous Chaos; Two Terms in the K-L Expansion.

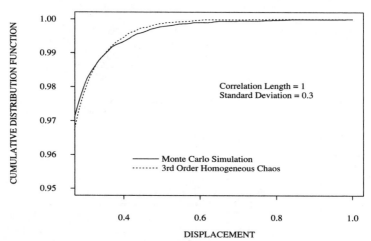

Figure 5.39:   Tail of the Cumulative Distribution Function for the Displacement at the Tip of the Beam Using 10,000-Sample Monte Carlo Simulation and a Third Order Homogeneous Chaos; Two Terms in the K-L Expansion.

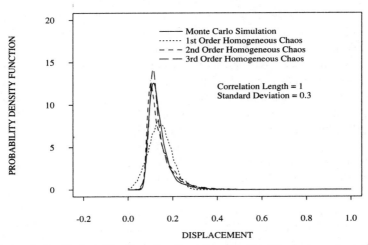

Figure 5.40: Probability Density Function for the Displacement at the Tip of the Beam Using 10,000-Sample Monte Carlo Simulation and up to Third Order Homogeneous Chaos; Four Terms in the K-L Expansion.

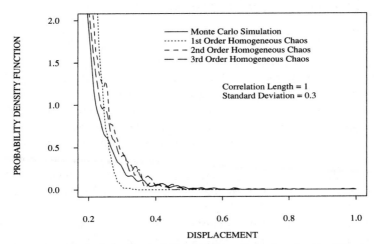

Figure 5.41: Tail of the Probability Density Function for the Displacement at the Tip of the Beam Using 10,000-Sample Monte Carlo Simulation and up to Third Order Homogeneous Chaos; Four Terms in the K-L Expansion.

yielding a first order filter. This agrees with the fact that the process being simulated, which corresponds to the covariance function given by equation (2.39), is a first order Markov process. The filter is then used to simulate five thousands realizations of the bending rigidity. For each of the simulated realizations, the associated deterministic problem is solved and a data bank for the response process is generated to compute its statistics. The process is repeated for various values of the standard deviation of the bending rigidity σ_{EI}. For σ_{EI} in the neighborhood of 0.3, the problem of dealing with realizations involving negative values of the bending rigidity becomes crucial. This problem is a consequence of the Gaussian assumption which permits negative values for the process EI. During the simulation, the small fraction of the realizations with negative EI is excluded. In essence, a truncated Gaussian distribution is used for EI to the extent that equation (2.20) is only approximately correct for the corresponding ξ_i. In physical problems of related interest, the coefficient of variation corresponding to a medium is usually less than 0.2, so that the Gaussian assumption is even more appropriate. The results from the simulation are shown on the same plots as the results obtained from the analytical analysis.

5.3   Two Dimensional Static Problem

5.3.1   Formulation

In this example, first a thin square plate is considered that is clamped along one edge and is subjected to a uniform in-plane tension along its opposite edge. The problem is depicted in Figure (5.44). The modulus of elasticity of the plate is assumed to be the realization of a two-dimensional Gaussian random process with known mean value $\bar{E}$ and known covariance function $C(\mathbf{x}_1, \mathbf{x}_2)$. Further, it is assumed that the external excitation is deterministic and of unit magnitude. The same problem is then formulated and solved for the plate with curved geometry shown in Figure (5.45), thus demonstrating the applicability of this stochastic finite element method to problems with arbitrary geometry.

In treating this problem a slightly different variational formulation is followed than the one used in the previous example. This is done to demonstrate the flexibility of the Karhunen-Loeve expansion, and its compatibility with various formulations of the finite element method. Let the domain A of the plate be discretized into N four-noded quadrilateral finite elements,

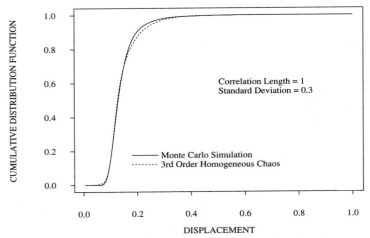

Figure 5.42: Cumulative Distribution Function for the Displacement at the Tip of the Beam Using 10,000-Sample Monte Carlo Simulation and a Third Order Homogeneous Chaos. Four Terms in the K-L Expansion.

each element having eight degrees of freedom. The strain energy V^e stored in each element A^e can be expressed as

$$V^e = \frac{1}{2} \int_{A^e} \boldsymbol{\sigma}^T(\mathbf{x}) \, \boldsymbol{\epsilon}(\mathbf{x}) \, dA^e \,, \tag{5.30}$$

where dA^e is a differential element in A^e. Further, $\boldsymbol{\sigma}(\mathbf{x})$ and $\boldsymbol{\epsilon}(\mathbf{x})$ denote the stress and the strain vectors respectively, as a function of the location $\mathbf{x}$ within each element. Assuming linear elastic material behavior, the stress may be expressed in terms of the strain as

$$\boldsymbol{\sigma} = \mathbf{D}^e \, \boldsymbol{\epsilon} \tag{5.31}$$

where $\mathbf{D}^e$ is the matrix of constitutive relations. Here $\boldsymbol{\sigma}$ and $\boldsymbol{\epsilon}$ are the vectors of stress and strain as given by the equation

$$\boldsymbol{\sigma}^T = \left[\, \sigma_{x_1} \; \sigma_{x_2} \; \tau_{x_1 x_2} \, \right] \tag{5.32}$$

$$\boldsymbol{\epsilon}^T = \left[\, \epsilon_{x_1} \; \epsilon_{x_2} \; \gamma_{x_1 x_2} \, \right] \,, \tag{5.33}$$

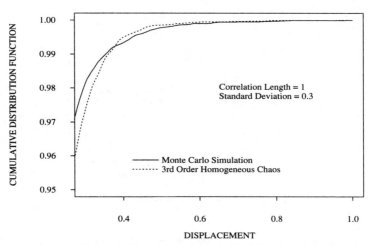

Figure 5.43:   Tail of the Cumulative Distribution Function for the Displacement at the Tip of the Beam Using 10,000-Sample Monte Carlo Simulation and a Third Order Homogeneous Chaos. Four Terms in the K-L Expansion.

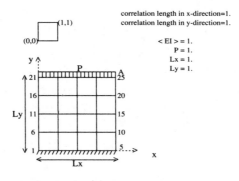

Figure 5.44: Plate with Random Rigidity; Exponential Covariance Model.

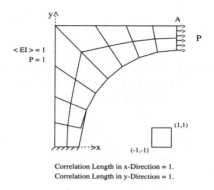

Correlation Length in x-Direction = 1.
Correlation Length in y-Direction = 1.

Figure 5.45: Plate with Random Rigidity, Exponential Covariance Model.

where σ_{x_i} is the stress along direction x_i and ϵ_{x_i} is the strain along that same direction. For the plane stress problem considered herein, $\mathbf{D}^e$ is given by the equation

$$\mathbf{D}^e = \frac{E^e(\mathbf{x}, \theta)}{1 - \mu_e^2} \left[\begin{array}{ccc} 1 & \mu_e & 0 \\ \mu_e & 1 & 0 \\ 0 & 0 & \frac{(1-\mu_e)}{2} \end{array} \right] = E^e(\mathbf{x}, \theta) \, \mathbf{P}^e , \qquad (5.34)$$

where $\mathbf{P}^e$ is a deterministic matrix, μ_e is the elemental Poisson ratio, and $E^e(\mathbf{x}, \theta)$ is the elemental modulus of elasticity. The two dimensional displacement vector $\mathbf{u}(\mathbf{x}, \theta)$ representing the longitudinal and transverse displacements within each element may be expressed in terms of the nodal displacements of the element in the form

$$\mathbf{u}(\mathbf{x}, \theta) = \mathbf{H}^e(r_1, r_2) \, \mathbf{U}^e(\theta) , \qquad (5.35)$$

where $\mathbf{H}^e(r_1, r_2)$ is the local interpolation matrix, $\mathbf{U}^e(\theta)$ is the random nodal response vector, and r_1 and r_2 are local coordinates over the element. For the rectangular elements used herein,

$$r_1 = \frac{(x_1 - x_{1c})}{l_{x_1}} , \; r_2 = \frac{(x_2 - x_{2c})}{l_{x_2}} . \tag{5.36}$$

Further, l_{x_i} is the dimension of the side of the rectangular element along the ith direction, and x_{ic} is the coordinate along that direction of the lower left corner of the element. The interpolation matrix $\mathbf{H}^e(r_1, r_2)$ can be expressed as

$$\mathbf{H}^e(r_1, r_2) = \left[\begin{array}{cccccccc} N_1 & 0 & N_2 & 0 & N_3 & 0 & N_4 & 0 \\ 0 & N_1 & 0 & N_2 & 0 & N_3 & 0 & N_4 \end{array} \right] , \tag{5.37}$$

where

$$N_1 = r_1(1 - r_2), \; N_2 = r_2(1 - r_1), \; N_3 = r_1 r_2, \; N_4 = (1 - r_1)(1 - r_2) . \tag{5.38}$$

Substituting equation (5.34) into equation (5.30) gives

$$V^e = \frac{1}{2} \int_{A^e} E^e(\mathbf{x}, \theta) \; \boldsymbol{\epsilon}^T(\mathbf{x}, \theta) \; \mathbf{P}^e \; \boldsymbol{\epsilon}(\mathbf{x}, \theta) \; dA^e . \tag{5.39}$$

The strain within an element is related to the displacements, longitudinal and transverse, through the relation

$$\boldsymbol{\epsilon}(\mathbf{x}, \theta) = \left[\begin{array}{cc} \dfrac{\partial}{\partial x_1} & 0 \\[2ex] 0 & \dfrac{\partial}{\partial x_2} \\[2ex] \dfrac{\partial}{\partial x_2} & \dfrac{\partial}{\partial x_1} \end{array} \right] \mathbf{u}(\mathbf{x}, \theta) . \tag{5.40}$$

Using equation (5.35), equation (5.40) is rewritten as

$$\boldsymbol{\epsilon}(\mathbf{x}, \theta) = \mathbf{B}^e \; \mathbf{U}^e , \tag{5.41}$$

with

$$
\mathbf{B}^e =
\begin{bmatrix}
\frac{(-1+r_2)}{l_{x_1}} & 0 & \frac{(1+r_2)}{l_{x_1}} & 0 & \frac{r_2}{l_{x_1}} & 0 & -\frac{r_2}{l_{x_1}} & 0 \\[2mm]
0 & \frac{(-1+r_1)}{l_{x_2}} & 0 & -\frac{r_1}{l_{x_2}} & 0 & \frac{r_1}{l_{x_2}} & 0 & \frac{(1-r_1)}{l_{x_2}} \\[2mm]
\frac{(-1+r_1)}{l_{x_2}} & \frac{(-1+r_2)}{l_{x_1}} & -\frac{r_1}{l_{x_2}} & \frac{(1+r_2)}{l_{x_1}} & \frac{r_1}{l_{x_2}} & \frac{r_2}{l_{x_1}} & \frac{1-r_1}{l_{x_2}} & -\frac{r_2}{l_{x_1}}
\end{bmatrix} .
$$

$$(5.42)$$

Substituting equation (5.41) back into equation (5.39) leads to

$$
V^e = \frac{1}{2} \mathbf{U}^{eT} \int_0^1 \int_0^1 E(r_1, r_2, \theta) \, \mathbf{B}^{eT}(r_1, r_2) \, \mathbf{P}^e \, \mathbf{B}^e(r_1, r_2) \, |\mathbf{J}^e| dr_1 \, dr_2 \, \mathbf{U}^e ,
$$

$$(5.43)$$

where $|\mathbf{J}^e|$ denotes the determinant of the Jacobian of the transformation that maps an arbitrary element (e) onto the four-noded square with sides equal to one. The integration is then performed over this *master* element. In performing the numerical integration, a four-point quadrature rule is used, and a procedure identical to the one described by Akin (1982) is employed to compute the value of the Jacobian at each integration point. The total strain energy V is obtained by summing the contributions from all the elements. This procedure gives

$$
V = \frac{1}{2} \sum_{e=1}^{N} \mathbf{U}^{eT} \int_0^1 \int_0^1 E(r_1, r_2, \theta) \, \mathbf{B}^{eT}(r_1, r_2) \, \mathbf{P}^e \, \mathbf{B}^e(r_1, r_2) \, |\mathbf{J}^e| \, dr_1 \, dr_2 \, \mathbf{U}^e .
$$

$$(5.44)$$

The local representation of the response is related to the global representation through the following transformation

$$
\mathbf{U}^e = \mathbf{C}^e \, \mathbf{U} ,
$$

$$(5.45)$$

where $\mathbf{C}^e$ is a rectangular permutation matrix of zeros and ones reflecting the connectivity of the elements and the topology of the mesh. Using equation (5.45), the following expression for the total energy stored in the system is obtained

$$
V = \frac{1}{2} \mathbf{U}^T \mathbf{K} \mathbf{U} ,
$$

$$(5.46)$$

where

$$\mathbf{K} = \sum_{e=1}^{N} \mathbf{C}^{eT} \mathbf{K}^e \mathbf{C}^e . \tag{5.47}$$

Equation (5.47) may be performed without computing the matrices $\mathbf{C}^e$, by using a bookkeeping procedure, as described by Akin (1982). The next stage in the computations involves solving the integral eigenvalue problem associated with the covariance kernel. That is,

$$\lambda_n f_n(x_1, y_1) = \int_A C(x_1, y_1; x_2, y_2) f_n(x_2, y_2) \, dx_1 \, dy_1 . \tag{5.48}$$

The kernel used in this example is defined by the equation

$$C(x_1, x_2; y_1, y_2) = e^{-|x_1 - y_1|/b_1 - |x_2 - y_2|/b_2} , \tag{5.49}$$

where b_1 and b_2 are the correlation distances in the x_1 and x_2 directions respectively.

Square Plate - Analytical Solution

Substituting equation (5.49) for the covariance kernel, and assuming that

$$f_n(x_1, x_2) = f_i^{(1)}(x_1) \, f_j^{(2)}(x_2) \tag{5.50}$$

and

$$\lambda_n = \lambda_i^{(1)} \lambda_j^{(2)} , \tag{5.51}$$

the solution of equation (5.48) reduces to the solution of the equation

$$\lambda_i^{(1)} \lambda_j^{(2)} f_i^{(1)}(x_1) f_j^{(2)}(x_2) = \tag{5.52}$$

$$\int_{-l_{x_1}/2}^{l_{x_1}/2} e^{-c_1|x_1 - y_1|} f_i^{(1)}(y_1) \, dy_1 \int_{-l_{x_2}/2}^{l_{x_2}/2} e^{--c_2|x_2 - y_2|} f_j^{(2)}(y_2) \, dy_2 .$$

where

$$c_i = \frac{1}{b_i} . \tag{5.53}$$

The solution of equation (5.52) is the product of the individual solutions of
the two equations

$$\lambda_i^{(1)} f_i^{(1)}(x_1) = \int_{-l_{x_1}/2}^{l_{x_1}/2} e^{-c_1|x_1-y_1|} f_i^{(1)}(y_1) \, dy_1 \qquad (5.54)$$

and

$$\lambda_j^{(2)} f_j^{(2)}(x_2) = \int_{-l_{x_2}/2}^{l_{x_2}/2} e^{-c_2|x_2-y_2|} f_j^{(2)}(y_2) \, dy_2 \ . \qquad (5.55)$$

Differentiating each of equations (5.54) and (5.55) twice, two second order
differential equations are obtained along with their associated boundary
conditions. The solution of the first of these equations produces the following
eigenvalues and normalized eigenfunctions,

$$\lambda_i^{(1)} = \frac{2\,c_1}{(\omega_i{}^2 + c_1^2)} \ , \qquad (5.56)$$

and

$$f_i^{(1)}(x) = \frac{\cos(\omega_i x)}{\sqrt{a + \dfrac{\sin(2\omega_i a)}{2\omega_i}}} \qquad for \ i \ odd \ , \qquad (5.57)$$

$$f_i^{(1)}(x) = \frac{\sin(\omega_i x)}{\sqrt{a - \dfrac{\sin(2\omega_i a)}{2\omega_i}}} \qquad for \ i \ even \ , \qquad (5.58)$$

In equations (5.56)-(5.58), the symbol ω_i refers to the solution of the following transcendental equation

$$\begin{cases} c_1 - \omega_i \tan(\omega_i a) = 0 \quad for \ i \ odd \\[2mm] and \\[2mm] \omega_i + c_1 \tan(\omega_i a) = 0 \quad for \ i \ even \ . \end{cases} \qquad (5.59)$$

The solution of equation (5.55) is identical to equations (5.56)-(5.59), with

c_1 replaced by c_2. Note that if $c_1 = c_2$, then to each eigenvalue there correspond two eigenfunctions of the form given by equation (5.50). The second function being obtained from the first one by permuting the subscripts. In this case, therefore, the complete normalized eigenfunctions are given by the equation

$$f_n(x,y) = \frac{1}{\sqrt{2}} \left[f_i^{(1)}(x)f_j^{(2)}(y) + f_j^{(1)}(x)f_i^{(2)}(y) \right] \qquad (5.60)$$

In the expansion of the random process, the terms are ordered in descending order of the magnitude of the eigenvalues λ_n.

Curved Plate - Numerical Solution

Subdividing the domain A of the plate into N finite elements A^e, equation (5.48) becomes

$$\lambda_n\, f_n(x_1,y_1) = \sum_{e=1}^{N} \int_{A^e} C(x_1,y_1;x_2,y_2)\, f_n(x_2,y_2)\, dA^e . \qquad (5.61)$$

Interpolating for the value of the unknown function within an element in terms of its nodal values results in the following expression

$$f_n(x,y) = \mathbf{H}^e(r_1,r_2)\, \mathbf{f}_n^e , \qquad (5.62)$$

where $\mathbf{H}^e(r_1,r_2)$ is the interpolation matrix in terms of local coordinates r_1 and r_2, and $\mathbf{f}_n^e$ is the vector of nodal values for the unknown function associated with element (e). For this particular problem, bilinear interpolation is used over four-noded quadrilateral element. The matrix $\mathbf{H}^e(r_1,r_2)$ is then given by the equation

$$\mathbf{H}^e(r_1,r_2) = \frac{1}{4} \left[\ (1-r_1)(1-r_2) \quad (1+r_1)(1-r_2) \right. \qquad (5.63)$$
$$\left. (1+r_1)(1+r_2) \quad (1-r_1)(1+r_2) \ \right] .$$

Substituting equation (5.62) and performing a transformation from global to local coordinates, equation (5.61) becomes

$$\lambda_n\, f_n(x_1,y_1) = \sum_{e=1}^{N} \int_{A^e} C(x_1,y_1;x_2,y_2)\, \mathbf{H}^e(r_1,r_2)\, |\mathbf{J}^e|\, dA^e\, \mathbf{f}_n^e , \qquad (5.64)$$

where $|\mathbf{J}^e|$ is the Jacobian of the coordinate transformation. A system of algebraic equations is obtained from equation (5.64) by requiring the corresponding error to be orthogonal to all the interpolation functions used. The result is an equation identical to equation (2.82),

$$\mathbf{C\,D} = \mathbf{\Lambda\,B\,D} , \qquad (5.65)$$

where now the jth column of $\mathbf{D}$ is the jth eigenfunction calculated at the nodal points and

$$\mathbf{\Lambda}_{ij} = \delta_{ij}\,\lambda_i . \qquad (5.66)$$

Matrices $\mathbf{C}$ and $\mathbf{B}$ are obtained by assembling matrices $\mathbf{C}_{ef}$ and $\mathbf{B}_{ef}$ where

$$\mathbf{C}_{ef} = \int_{A^e} \int_{A^f} C(x_1, y_1; x_2, y_2)\,\mathbf{H}^{eT}(r_1, r_2)\,\mathbf{H}^f(r_1, r_2)\,dA^e\,dA^f , \qquad (5.67)$$

and

$$\mathbf{B}_{ef} = \int_{A^e} C(x_1, y_1; x_2, y_2)\,\mathbf{H}^{eT}(r_1, r_2)\,\mathbf{H}^f(r_1, r_2)\,dA^e , \qquad (5.68)$$

where (x_1, y_1) and (r_1, r_2) denote respectively the global and local coordinates of a point in A^e. The assembly procedure just mentioned consists of combining entries corresponding to the same node (Akin, 1982). Once the eigenvectors and eigenfunctions have been calculated, the solution procedure is similar to the one employed for the square plate example, and the two will be combined from here on.

The Karhunen-Loeve expansion for the modulus of elasticity may be substituted into equation (5.44) to transform equation (5.46) into

$$V = \frac{1}{2}\,\mathbf{U}^T \sum_{k=1}^{M} \xi_k(\theta)\,\mathbf{K}^{(k)}\,\mathbf{U} . \qquad (5.69)$$

The integrations indicated in equation (5.44) may be performed either analytically or using some numerical quadrature scheme. The work performed by the externally applied forces is

$$V' = \sum_{e=1}^{N} \int_{A^e} \mathbf{u}^{e\,T}(\mathbf{x}, \theta)\,\mathbf{f}(\mathbf{x})\,dA^e ,$$

$$= \mathbf{U}^T \sum_{k=1}^{M} \mathbf{C}^{eT} \int_{A^e} \mathbf{H}^{e\,T}(\mathbf{x})\,\mathbf{f}(\mathbf{x})\,dA^e = \mathbf{U}^T\,\mathbf{f} , \qquad (5.70)$$

where

$$\mathbf{f} = \sum_{e=1}^{N} \mathbf{C}^{eT} \int_{A^e} \mathbf{H}^e(\mathbf{x}) \, \mathbf{f}(\mathbf{x}) \, dA^e \; . \tag{5.71}$$

Minimizing the total potential energy $(V - V')$ with respect to $\mathbf{U}$, leads to the equation

$$\left[\mathbf{K}^{(0)} + \sum_{k=1}^{M} \xi_k(\theta) \, \mathbf{K}^{(k)} \right] \mathbf{U} = \mathbf{f} \; . \tag{5.72}$$

From this point on, the treatment of this two dimensional problem is identical to that for the previous one-dimensional beam problem. Specifically, the response vector $\mathbf{U}$ is expanded as

$$\mathbf{U} = \sum_{i=0}^{P} \mathbf{c}_i \, \Psi_i[\{\xi_l\}] \; , \tag{5.73}$$

with the vector coefficients $\mathbf{c}_i$ evaluated as the solution to the following system of algebraic equations

$$\left[\sum_{k=0}^{M} <\xi_k \Psi_i[\{\xi_l\}] \Psi_j[\{\xi_l\}]> \, \mathbf{K}^{(k)} \right] \mathbf{c}_i = \; <\Psi_j[\{\xi_l\}] \, \mathbf{f}> \; , \quad j = 1 \ldots P \; . \tag{5.74}$$

5.3.2   Results

Analytical Solution

The square plate is divided into sixteen elements, four along each direction. The nodes are numbered in increasing order away from the clamped edge and counting from left to right. Figures (5.46) and (5.47) show the standard deviation of the response at the free corner of the plate versus that of the modulus of elasticity for the improved Neumann expansion and for the Homogeneous Chaos expansion, respectively. Observe that better convergence is achieved with a third order Homogeneous Chaos expansion than with a fourth order Neumann expansion. Figures (5.48)-(5.50) show the projections on the Homogeneous Chaos for various orders of the expansion. These are the vectors $\mathbf{c}_i$ appearing in equation (5.74). Note that the fluctuations of the nodal values of these vectors reflects the fact that the two-dimensional geometry of the plate is represented by a one-dimensional figure. Further,

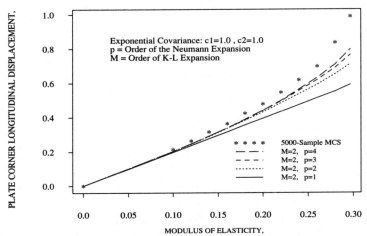

Figure 5.46: Normalized Standard Deviation of Longitudinal Displacement at Corner A of the Rectangular Plate, versus Standard Deviation of the Modulus of Elasticity; $\sigma_{max} = 0.433$; Exponential Covariance; Neumann Expansion Solution.

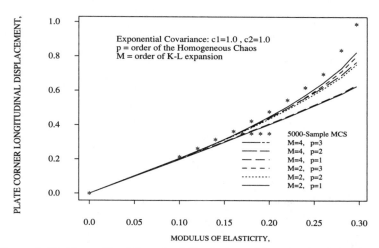

Figure 5.47: Normalized Standard Deviation of Longitudinal Displacement at Corner A of the Rectangular Plate, versus Standard Deviation of the Modulus of Elasticity; $\sigma_{max} = 0.433$; Exponential Covariance; Polynomial Chaos Solution

it is seen that the third order Chaos contributes a small amount to the total variation as compared to the first two. Figures (5.51)-(5.55) show the convergence of the individual projections as the order of the expansion is increased, that is as the value of P in equation (5.73) is increased. Figures (5.56)-(5.63) show the results corresponding to the probability distribution of one of the response variable at the free corner of the plate. The method described in section (4.4) is used here again.

The curved plate is shown in Figure (5.45). The curved side is a ninety degree arc of a circle of unit radius. The length of the straight edges is equal to 1.25. The standard deviation of the longitudinal and transverse displacements at node A, using two terms in the K-L expansion for the material stochasticity, is shown in Figures (5.64) and (5.65), plotted against the standard deviation of the modulus of elasticity. The results corresponding to four and six terms in the K-L expansion are shown in Figures (5.66)-(5.67), and (5.68)-(5.69), respectively. Note the excellent convergence. Also note that for this example, the effect of the number of terms used in the Karhunen-Loeve expansion of the material stochasticity is more pronounced than for

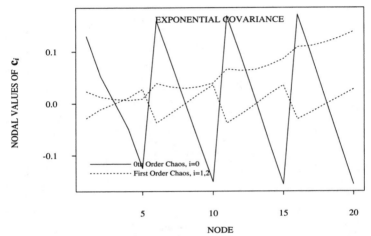

Figure 5.48: Linear Interpolation of the Nodal Values of the Vector c_i of Equation (5.73) for the Rectangular Plate Stretching Problem, $i = 0, 1, 2$; Longitudinal Displacement Representation; 2 Terms in K-L Expansion, $M = 2$; $p = 0, 1$.

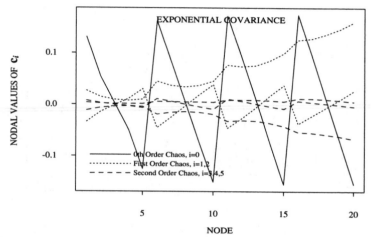

Figure 5.49: Linear Interpolation of the Nodal Values of the Vector $\mathbf{c}_i$ of Equation (5.73) for the Rectangular Plate Stretching Problem, $i = 0, ..., 5$; Longitudinal Displacement Representation; 2 Terms in K-L Expansion, $M = 2$; $p = 0, 1, 2$.

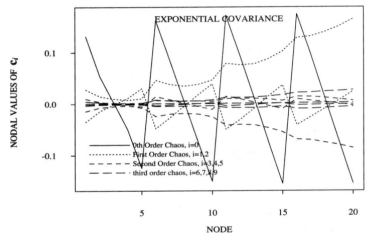

Figure 5.50: Linear Interpolation of the Nodal Values of the Vector $\mathbf{c}_i$ of Equation (5.73) for the Rectangular Plate Stretching Problem, $i = 0, ..., 9$; Longitudinal Displacement Representation; 2 Terms in K-L Expansion, $M = 2$; $p = 0, 1, 2, 3$.

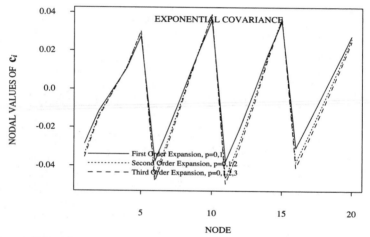

Figure 5.51: Linear Interpolation of the Nodal Values of the Vector c_i of Equation (5.73) for the Rectangular Plate Stretching Problem, $i = 1$; Longitudinal Displacement Representation; 2 Terms in K-L Expansion, $M = 2$; $P = 3, 6, 10$.

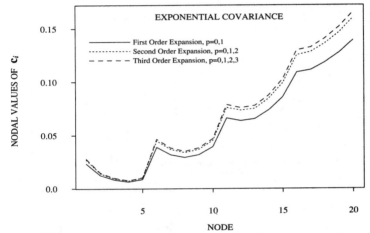

Figure 5.52: Linear Interpolation of the Nodal Values of the Vector c_i of Equation (5.73) for the Rectangular Plate Stretching Problem, $i = 2$; Longitudinal Displacement Representation; 2 Terms in K-L Expansion, $M = 2$; $P = 3, 6, 10$.

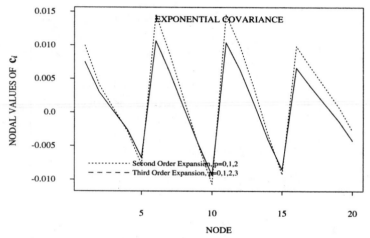

Figure 5.53: Linear Interpolation of the Nodal Values of the Vector c_i of Equation (5.73) for the Rectangular Plate Stretching Problem, $i = 3$; Longitudinal Displacement Representation; 2 Terms in K-L Expansion, $M = 2$; $P = 6, 10$.

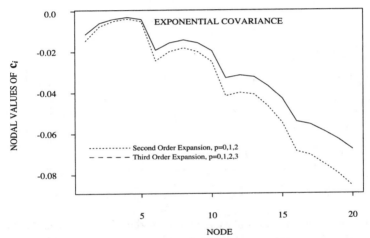

Figure 5.54: Linear Interpolation of the Nodal Values of the Vector $\mathbf{c}_i$ of Equation (5.73) for the Rectangular Plate Stretching Problem, $i = 4$; Longitudinal Displacement Representation; 2 Terms in K-L Expansion, $M = 2$; $P = 6, 10$.

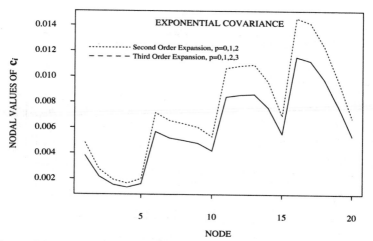

Figure 5.55: Linear Interpolation of the Nodal Values of the Vector $\mathbf{c}_i$ of Equation (5.73) for the Rectangular Plate Stretching Problem, $i = 5$; Longitudinal Displacement Representation; 2 Terms in K-L Expansion, $M = 2$; $P = 6, 10$.

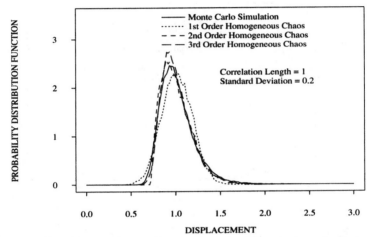

Figure 5.56: Longitudinal Displacement at the Free End of the Rectangular Plate; Probability Density Function Using 30,000-Sample MSC, and Using Third Order Homogeneous Chaos; Two Terms in the K-L Expansion; Exponential Covariance.

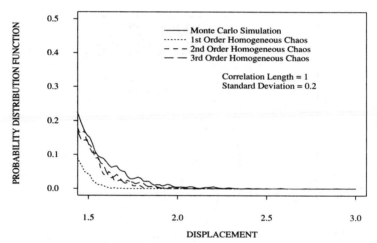

Figure 5.57: Longitudinal Displacement at the Free End of the Rectangular Plate; Tail of the Probability Density Function Using 30,000-Sample MSC, and Using Third Order Homogeneous Chaos; Two Terms in the K-L Expansion; Exponential Covariance.

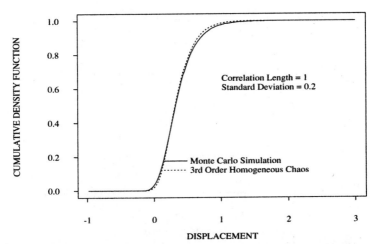

Figure 5.58: Longitudinal Displacement at the Free End of the Rectangular Plate; Cumulative Distribution Function Using 30,000-Sample MSC, and Using Third Order Homogeneous Chaos; Two Terms in the K-L Expansion; Exponential Covariance.

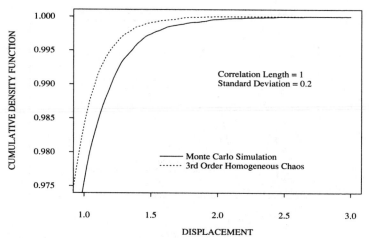

Figure 5.59: Longitudinal Displacement at the Free End of the Rectangular Plate; Tail of the Cumulative Distribution Function Using 30,000-Sample MSC, and Using Third Order Homogeneous Chaos; Two Terms in the K-L Expansion; Exponential Covariance.

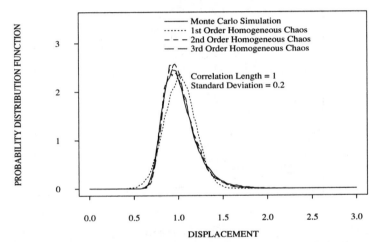

Figure 5.60: Longitudinal Displacement at the Free End of the Rectangular Plate; Probability Density Function Using 30,000-Sample MSC, and Using Third Order Homogeneous Chaos; Four Terms in the K-L Expansion; Exponential Covariance.

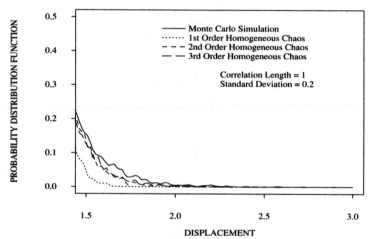

Figure 5.61: Longitudinal Displacement at the Free End of the Rectangular Plate; Tail of the Probability Density Function Using 30,000-Sample MSC, and Using Third Order Homogeneous Chaos; Four Terms in the K-L Expansion; Exponential Covariance.

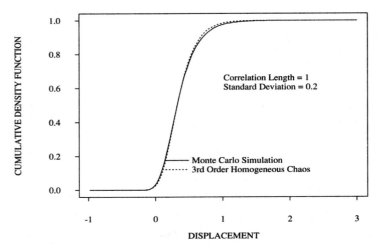

Figure 5.62: Longitudinal Displacement at the Free End of the Rectangular Plate; Cumulative Distribution Function Using 30,000-Sample MSC, and Using Third Order Homogeneous Chaos; Four Terms in the K-L Expansion; Exponential Covariance.

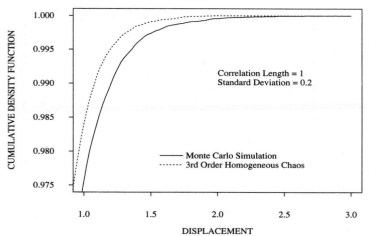

Figure 5.63: Longitudinal Displacement at the Free End of the Rectangular Plate; Tail of the Cumulative Distribution Function Using 30,000-Sample MSC, and Using Third Order Homogeneous Chaos; Four Terms in the K-L Expansion; Exponential Covariance.

the square plate example. This behavior is attributed to the fact that in this case the random field involves intricate geometric boundaries; therefore more terms are required for its adequate representation. The probability distribution functions corresponding to one of the response variable at node A are depicted in Figures (5.70)-(5.75). Results corresponding to up to six terms in the Karhunen-Loeve expansion and third order Polynomial Chaos are shown.

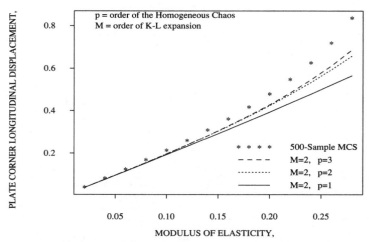

Figure 5.64: Normalized Standard Deviation of Longitudinal Displacement at Corner A of the Curved Plate, versus Standard Deviation of the Modulus of Elasticity; Two Terms in the K-L Expansion; Exponential Covariance; $\sigma_{max} = 19.4$.

Monte Carlo Simulation

The two dimensional process representing the modulus of elasticity of the plate is simulated as follows. First, the covariance matrix is constructed, the ij^{th} element of which corresponds to the correlation of the process at points i and j. Following that, the Cholesky decomposition of the positive definite covariance matrix is obtained (Golub and Van Loan, 1984). The columns of the Cholesky decomposition factor are then used as the basis for simulating the process. In other words, the process is obtained by premulti-

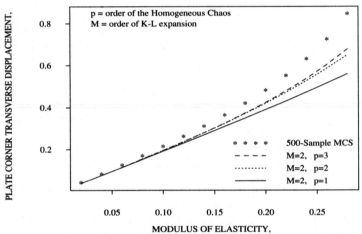

Figure 5.65: Normalized Standard Deviation of Transverse Displacement at Corner A of the Curved Plate, versus Standard Deviation of the Modulus of Elasticity; Two Terms in the K-L Expansion; Exponential Covariance; $\sigma_{max} = 20.4$.

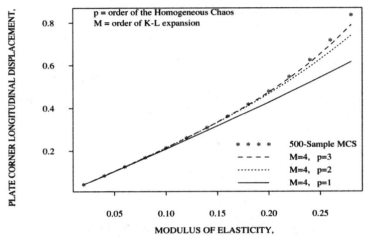

Figure 5.66: Normalized Standard Deviation of Longitudinal Displacement at Corner A of the Curved Plate, versus Standard Deviation of the Modulus of Elasticity; Four Terms in the K-L Expansion; Exponential Covariance; $\sigma_{max} = 19.4$.

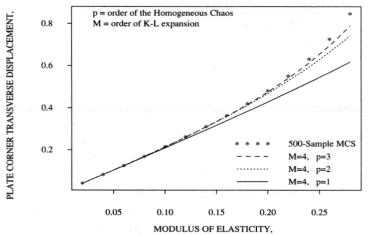

Figure 5.67: Normalized Standard Deviation of Transverse Displacement at Corner A of the Curved Plate, versus Standard Deviation of the Modulus of Elasticity; Four Terms in the K-L Expansion; Exponential Covariance; $\sigma_{max} = 20.4$.

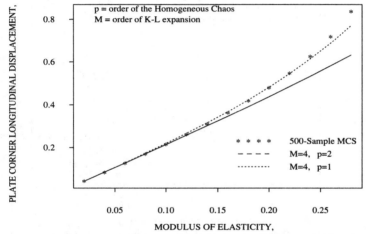

Figure 5.68: Normalized Standard Deviation of Longitudinal Displacement at Corner A of the Curved Plate, versus Standard Deviation of the Modulus of Elasticity; Six Terms in the K-L Expansion; Exponential Covariance; $\sigma_{max} = 19.4$.

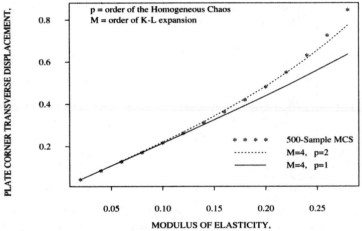

Figure 5.69: Normalized Standard Deviation of Transverse Displacement at Corner A of the Curved Plate, versus Standard Deviation of the Modulus of Elasticity; Six Terms in the K-L Expansion; Exponential Covariance; $\sigma_{max} = 20.4$.

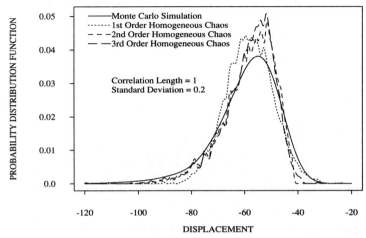

Figure 5.70: Longitudinal Displacement at the Free End of the Curved Plate; Probability Density Function Using 30,000-Samples MSC, and Using Third Order Homogeneous Chaos; Two Terms in the K-L Expansion; Exponential Covariance.

plying the Cholesky factor with a vector consisting of Gaussian uncorrelated variates. For the example involving the curved geometry problem, a different simulation technique is used. It is prompted by the fact that the nodal points are not uniformly distributed over the domain of the plate. The issue is addressed by using the Karhunen-Loeve expansion to simulate a truly continuous random field. To this end, the eigenvalues and eigenfunctions of the covariance kernel are computed as described in the previous section. The random field is then simulated using equation (2.6) with the number of terms equal to the number of nodes in the system. The orthogonal random variables appearing in that equation are obtained as pseudorandom computer generated uncorrelated variates, with zero mean and unit variance. The resulting simulated random field is not as sensitive to the mesh size and nodal point distribution as the field obtained using the more conventional procedure described for the previous example. The results from using the Monte Carlo simulation method are superimposed on the same plot as the analytical results. Observe the good agreement between the analytical and the simulated results even for large values of the coefficient of variation.

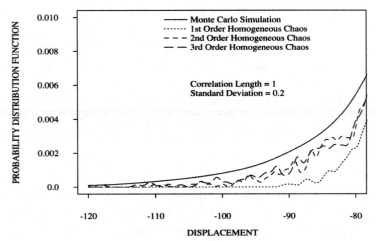

Figure 5.71: Longitudinal Displacement at the Free End of the Curved Plate; Tail of the Probability Density Function Using 30,000-Sample MSC, and Using Third Order Homogeneous Chaos; Two Terms in the K-L Expansion; Exponential Covariance.

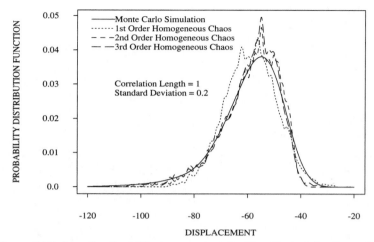

Figure 5.72: Longitudinal Displacement at the Free End of the Curved Plate; Probability Density Function Using 30,000-Sample MSC, and Using Third Order Homogeneous Chaos; Four Terms in the K-L Expansion; Exponential Covariance.

5.4 One Dimensional Dynamic Problem

5.4.1 Description of the Problem

The last example to be considered involves an Euler-Bernoulli beam shown, in Figure (5.76), of length L, modulus of elasticity E, area moment of inertia I, and mass density m. Let the beam be supported by an elastic foundation having a reaction modulus $k(x, \theta)$ which is considered to be a one-dimensional Gaussian random process with mean value $\bar{k}$ and covariance function $C_{kk}(x, t, \theta)$. The beam is assumed to be acted upon by a zero-mean stationary random process $f(x, t, \theta)$, featuring both temporal and spatial random fluctuations, as reflected by its spectral density function which is given by the equation

$$S_{ff}(x_1, x_2; \omega) = e^{-|x_1 - x_2|/\omega b}, \quad |\omega b| > 0.1 \tag{5.75}$$

where x_1 and x_2 denote two locations on the beam, ω is the wave number associated with time, and b is the correlation length of the excitation process.

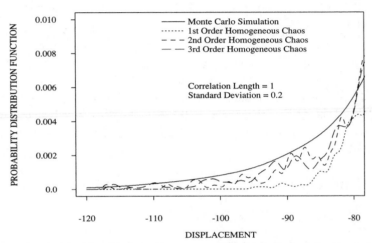

Figure 5.73: Longitudinal Displacement at the Free End of the Curved Plate; Tail of the Probability Density Function Using 30,000-Sample MSC, and Using Third Order Homogeneous Chaos; Four Terms in the K-L Expansion; Exponential Covariance.

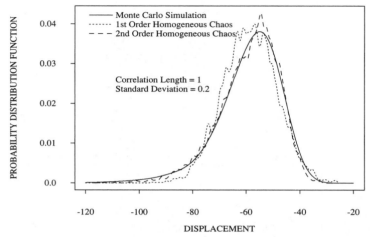

Figure 5.74: Longitudinal Displacement at the Free End of the Curved Plate; Probability Density Function Using 30,000-Sample MSC, and Using Third Order Homogeneous Chaos; Six Terms in the K-L Expansion; Exponential Covariance.

In the present analysis, the correlation length b is assumed to be a constant, although any dependence on frequency can be accommodated.

5.4.2 Implementation

The differential equation governing the motion of the beam with constant bending rigidity is

$$m \, \frac{\partial^2}{\partial t^2} u(x,t,\theta) \; + \; c\frac{\partial}{\partial t} u(x,t,\theta) \quad + EI \, \frac{\partial^4}{\partial x^4} u(x,t,\theta) \qquad (5.76)$$
$$+ \, k(x,\theta) \, u(x,t,\theta) \; = \; f(x,t,\theta) \, ,$$

where c is a coefficient of viscous damping. Adopting the discretization procedure of the previous examples, the beam may be divided into N finite elements. Then, an equation involving $2N \times 2N$ matrices is obtained of the form

$$\mathbf{M} \, \ddot{\mathbf{U}}(t,\theta) \; + \; \mathbf{C} \, \dot{\mathbf{U}}(t,\theta) \; + \; \mathbf{K} \, \mathbf{U}(t,\theta) \; + \; \mathbf{K}_f \, \mathbf{U}(t,\theta) \; = \; \mathbf{f}(t,\theta) \, , \quad (5.77)$$

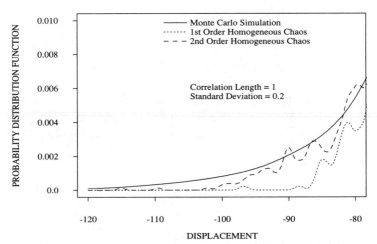

Figure 5.75: Longitudinal Displacement at the Free End of the Curved Plate; Tail of the Probability Density Function Using 30,000-Sample MSC, and Using Third Order Homogeneous Chaos; Six Terms in the K-L Expansion; Exponential Covariance.

where a dot denotes differentiation with respect to time and $\mathbf{M}$, $\mathbf{K}$, $\mathbf{K_f}$ are generated by assembling the elemental matrices

$$\mathbf{M}^e = \int_{l^e} \mathbf{H}^{eT} \, m \, \mathbf{H}^e \, dl^e \qquad (5.78)$$

$$\mathbf{K}^e = EI \int_{l^e} \mathbf{B}^{eT} \, \mathbf{B}^e \, dl^e \qquad (5.79)$$

and

$$\mathbf{K}_f^e = \int_{l^e} k(x,\theta) \, \mathbf{H}^{eT} \, \mathbf{H}^e \, dl^e \; . \qquad (5.80)$$

Further, the damping matrix $\mathbf{C}$ is assumed, for simplicity, to be of the form

$$\mathbf{C} = c_M \, \mathbf{M} + c_K \, \mathbf{K} \; , \qquad (5.81)$$

where c_M and c_K are constants.

Replacing the process $k(x,\theta)$ in equation (5.80) by its Karhunen-Loeve series gives

$$\mathbf{K}_f^e = \int_{l^e} \bar{k}(x) \, \mathbf{H}^{eT} \, \mathbf{H}^e \, dl^e \; + \; \sum_{k=1}^{M} \xi_k \int_{l^e} \sqrt{\lambda_k} f_k(x) \, \mathbf{H}^{eT} \, \mathbf{H}^e \, dl^e \; , \quad (5.82)$$

where $f_k(x)$ is the kth eigenfunction of the covariance kernel and λ_k is the corresponding k^{th} eigenvalue. Equation (5.77) then becomes,

$$\mathbf{M} \, \ddot{\mathbf{U}}(t) \; + \; \mathbf{C} \, \dot{\mathbf{U}}(t) \; + \; \mathbf{K} \, \mathbf{U}(t) \; + \; \bar{\mathbf{K}}_f \, \mathbf{U}(t) \qquad (5.83)$$

$$+ \; \sum_{k=1}^{M} \xi_k \mathbf{K}_f^{(k)} \, \mathbf{U}(t) \; = \; \mathbf{f}(t) \; ,$$

in which the argument θ is deleted for notational simplicity. Taking the Fourier transform of equation (5.83) yields

$$\left[-\omega^2 \mathbf{M} + i\omega \mathbf{C} + \mathbf{K} + \bar{\mathbf{K}}_{\mathbf{f}} + \sum_{k=1}^{M} \xi_k \mathbf{K}_f^{(k)} \right] \mathbf{U}(\omega) = \mathbf{F}(\omega), \qquad (5.84)$$

where

$$\mathbf{U}(\omega) = \int_{-\infty}^{\infty} \mathbf{U}(t) \, e^{-i\omega t} \, dt \; , \qquad (5.85)$$

and

$$\mathbf{F}(\omega) = \int_{-\infty}^{\infty} \mathbf{f}(t) \, e^{-i\omega t} \, dt \; . \tag{5.86}$$

Equation (5.84) may be rewritten as

$$\left[\mathbf{H}(\omega) + \sum_{k=1}^{M} \xi_k \, \mathbf{K}_f^{(k)} \right] \mathbf{U}(\omega) = \mathbf{F}(\omega) \; , \tag{5.87}$$

where

$$\mathbf{H}(\omega) = \left[-\omega^2 \, \mathbf{M} + i\omega \, \mathbf{C} + \mathbf{K} + \bar{\mathbf{K}}_{\mathbf{f}} \right] \tag{5.88}$$

is a deterministic matrix. Equivalently, equation (5.87) can be written as

$$\left[\mathbf{I} + \sum_{k=1}^{M} \xi_k \, \mathbf{Q}_f^{(k)} \right] \mathbf{U}(\omega) = \mathbf{P}(\omega) \; , \tag{5.89}$$

where

$$\mathbf{Q}_f^{(k)} = \mathbf{H}(\omega)^{-1} \, \mathbf{K}_f^{(k)} \; , \tag{5.90}$$

and

$$\mathbf{P}(\omega) = \mathbf{H}(\omega)^{-1} \, \mathbf{F}(\omega) \; . \tag{5.91}$$

Equation (5.89) has the same form as equation (5.28). The spectral density of the response process can be written as

$$\mathbf{S}_{UU}(\omega) = \left[\mathbf{I} + \sum_{k=1}^{M} \xi_k \, \mathbf{Q}_f^{(k)} \right]^{-1} \mathbf{S}_{PP}(\omega) \left[\mathbf{I} + \sum_{k=1}^{M} \xi_k \, \mathbf{Q}_f^{(k)} \right]^{-H} , \tag{5.92}$$

where the superscript H denotes hermitian transposition. Using the Neumann expansion for the inverse operator, equation (5.92) can be put in the form

$$\mathbf{S}_{UU}(\omega) = \sum_{i=1}^{\infty} \sum_{j=1}^{\infty} (-1)^{i+j} \left[\xi_k \, \mathbf{Q}_f^{(k)} \right]^{i} \mathbf{S}_{PP}(\omega) \left[\xi_k \, \mathbf{Q}_f^{(k)} \right]^{j} . \tag{5.93}$$

Alternatively, using the Homogeneous Chaos approach, the spectral density of the response can be represented in the following form

$$\mathbf{S}_{UU}(\omega) = \sum_{i=1}^{\infty} \sum_{j=1}^{\infty} \Gamma_i(\{\xi\}) \, \Gamma_j(\{\xi\}) \, \mathbf{a}_i \, S_{PP}(\omega) \, \mathbf{a}_j^T \qquad (5.94)$$

where $\mathbf{a}_i$ is a vector that is obtained as the solution to a system of linear equations as indicated earlier. Note that the spectral density matrix of $\mathbf{P}(\omega)$ is related to the spectral density matrix of $\mathbf{F}(\omega)$ by the equation

$$\mathbf{S}_{PP}(\omega) = \mathbf{H}(\omega)^{-1} \, \mathbf{S}_{ff}(\omega) \, \mathbf{H}(\omega)^{-H} \, . \qquad (5.95)$$

In addition, $S_{ff}(x_1, x_2; \omega)$ is given by equation (5.75) which may be regarded as a frequency dependent spatial correlation. Thus, it is numerically efficient to expand $S_{ff}(x_1, x_2; \omega)$ in its spectral series and perform the spatial discretization on the eigenfunctions of the expansion. Indeed, this procedure is adopted in the finite element code developed to solve this problem. Specifically, $S_{ff}(x_1, x_2; \omega)$ is expanded as

$$S_{ff}(x_1, x_2; \omega) = \sum_{i=1}^{K} \mu_i \, g_i(x_1; \omega) \, g_i(x_2; \omega) \, . \qquad (5.96)$$

In equation (5.96), μ_i refers to the eigenvalues associated with the kernel $S_{ff}(x_1, x_2; \omega)$, while $g_i(x; \omega)$ denote the corresponding eigenfunctions.

5.4.3   Results

In the numerical implementation of this problem, the exponential covariance kernel is used for the reaction modulus $k(x, \theta)$, with a correlation length equal to 1. The numerical values of other parameters are included in Figure (5.76). Figure (5.77) shows the spectral density $S_{uu}(\omega)$ of the displacement at one end of the beam along with pertinent Monte Carlo simulation results for a frequency increment equal to 0.1. Note the excellent agreement of the two solutions until the exact damped natural frequency of the beam is approached, causing the norm of the matrix $\left[I + \sum_{k=1}^{M} \xi_k \, \mathbf{Q}_f^{(k)} \right]$ to increase beyond the radius of convergence of the Neumann expansion. This

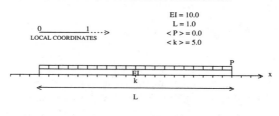

Figure 5.76: Beam on Random Elastic Foundation Subjected to a Random Dynamic Excitation; Exponential Covariance Model.

problem is not encountered with the Homogeneous Chaos approach, a consequence of the completeness of the Polynomial Chaos basis in the space of random variables. Figure (5.78) and (5.79) show the results corresponding to successive orders of approximation using the Improved Neumann expansion and the Homogeneous Chaos methods respectively. The Monte Carlo simulation is obtained, as described in the previous section, using the Cholesky decomposition of the covariance matrix. It is noted that using more than four terms in the spectral expansion of $S_{ff}(x_1, x_2; \omega)$, equation (5.96), does not improve significantly the quality of the derived solutions. Clearly, the number of terms needed for a particular problem depends on the magnitude of the correlation length of the excitation process.

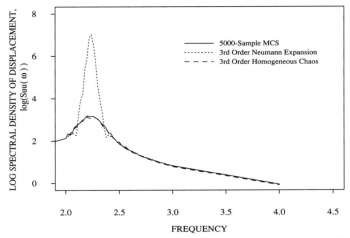

Figure 5.77: Spectral Density of the Displacement at the End of the Beam on Random Elastic Foundation.

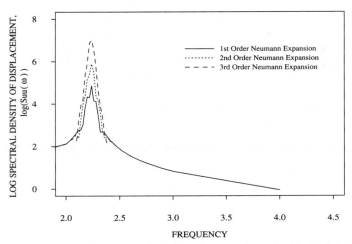

Figure 5.78: Spectral Density of the Displacement at the End of the Beam on Random Elastic Foundation.

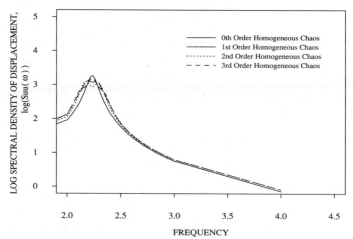

Figure 5.79: Spectral Density of the Displacement at the End of the Beam on Random Elastic Foundation.

Chapter 6

SUMMARY AND CONCLUDING REMARKS

6.1

This monograph has considered a class of engineering systems whose behavior is governed by linear differential equations with coefficients that are modeled as random processes. These coefficients can be associated with random variability in the material properties of the system. In this regard, a material property of an actual system can be thought of as a realization of an appropriate second-order stochastic process.

Emphasis has been given to a spectral approach to determining the system response. Specifically, the discussed solution approach involves two stages. The first stage hinges upon adequately representing the stochastic processes corresponding to the random system properties. This has been achieved by means of the Karhunen-Loeve expansion. This expansion provides an optimal mean-square convergent representation of a stochastic process in terms of a denumerable set of uncorrelated random variables; the expansion has the quite desirable feature of being valid over the entire domain of definition of the problem. The end product from the first stage is a matrix equation involving deterministic matrices multiplied by random variables. The second stage in the solution approach consists of solving the resulting set of equations. Two options have been presented to this end. The first one relies on utilizing the Neumann expansion for the inverse matrix to derive an explicit expression for the response process as a multidimensional homogeneous polynomial in the uncorrelated random variables. The second option consists of a Galerkin scheme whereby the solution process is approximated by its projection onto a complete basis in the space of random variables. Such a basis has been identified as consisting of the Polynomial Chaoses. These are orthogonal multi-dimensional polynomials in the uncorrelated random variables. An explicit expression for the response process is thus obtained. The response is represented as a surface in the multi-dimensional space defined by the uncorrelated random variables. This form for the solution process is particularly suitable for reliability analysis of the system being considered.

The spectral approach has been applied to typical problems of statics and dynamics. The obtained results have been found in good agreement with data generated from Monte Carlo simulation solutions. These results pertain to second-order statistics of the response process, as well as, to its probability density function. A somewhat detailed description of the content of each chapter follows.

In Chapter I the mathematical tools needed to adequately address the problem were presented. These consisted mainly of the concepts of Hilbert space bases and projections. The concept of a ring of functions that is dense in the Hilbert space of random functions was also introduced as providing the necessary mathematical background for the Polynomial Chaoses introduced in the sequel.

Chapter II dealt with the important issue of representing stochastic processes. The discussion was confined to continuous-parameter processes since they constitute the majority of mathematical models used for processes encountered in practice. Two major concepts in relation to the representation of stochastic processes were pointed out. First was the representation of a stochastic process by another stochastic process that is more tractable analytically. A relevant example is the spectral representation of stochastic processes in terms of a set of continuous orthogonal functions. Although this kind of representation is pivotal in the theoretical developments of analyzing systems with random parameters, it is of little use in a computational environment where only countable and discrete quantities can be implemented. Addressing this issue, another kind of representation of stochastic processes was introduced involving a denumerable set of random variables. This representation utilized expansions in terms of a finite dimensional basis in the space of random functions. In this connection a stochastic process was viewed as a curve in this Hilbert space. As a particularly important and useful example, the Karhunen-Loeve expansion was introduced as an optimal representation of this kind. The expansion was considered as a representation of the stochastic process along an orthogonal basis in the space of random variables in terms of the eigenfunctions of the covariance function of the process. This fact was used to view the expansion as an expression of a congruence between the Hilbert space spanned by the random variable basis and the Hilbert space defined by the covariance function. The Karhunen-Loeve expansion was shown to be well suited for computational purposes since it is optimal, convergent, and involves a denumerable set of random variables. Crucial to implementing the Karhunen-Loeve expansion is a specification of the covariance function of the process to be represented. Such information is known in relation to the processes describing the material properties, but is not available, in general, for the unknown response process. This issue was addressed by introducing the Homogeneous Chaos expansion. This is a mean-square convergent representation of multidimensional functions of Gaussian variables in terms of the Polynomial Chaoses that are orthogonal with respect to the Gaussian measure. It can be viewed as the discrete equivalent of the Wiener-Hermite expansion and may be obtained from the Cameron-Martin expansion by a limiting process. The successive Polynomial Chaoses were shown to be readily computable either algorithmically or with the use of symbolic computation programs. It was shown that any random function admits a mean-square convergent expansion along the basis provided by the Polynomial Chaoses. The importance of such a representation in connection with a finite element analysis is great. Furthermore, it was shown that using the Karhunen-Loeve expansion to represent the random system parameters circumvents the problem of making the size of the finite element mesh depend on the level of random fluctuations of these parameters. An abstract mesh was thus provided along the random dimension that was uncoupled from the physical finite element mesh. It was indicated that this abstract mesh is actually a discretization along the random dimension with respect to the spectral measure associated with the Polynomial Chaos basis.

In Chapter III the deterministic finite element method was reviewed first. Next, some of the difficulties associated with extending the analysis to stochastic systems were identified. The inadequacy of non-spectral techniques to effectively deal with these issues was also indicated. Specifically, the large number of random variables resulting from the pointwise representation of the stochastic processes involved, and the lack of a consistent rationale to invert the resultant random matrices were pointed out. In this context, perturbation methods and Neumann expansion methods were presented as the major tools available to address the issue. The methods have been proven successful in dealing with problems involving randomness of small magnitude. They have been found quite difficult to implement for sizable randomness requiring a significant number of terms to be included in the series representation of the solution process. In the remainder of the chapter, two other methods were discussed that overcome, to a large extent, the difficulties involved. Both methods rely on the Karhunen-Loeve expansion presented in Chapter III to recast the problem in terms of a denumerable set of uncorrelated random variables. When coupled with the K-L expansion, deterministic finite element methods provided matrix equations that were much simpler than the equations obtained by the other methods. As a first approach to solving the resulting equations, the Neumann expansion for the inverse was applied. The procedure was much easier to implement, having introduced the K-L expansion, than it would have been otherwise. This was due to the fact that the matrix to be inverted consisted of the sum of deterministic matrices with the random variables appearing only as multiplicative factors. This yielded a series expression for the solution process as a multidimensional homogeneous polynomial in the uncorrelated random variables. The second approach to solving the equations resulting from coupling the finite element discretization and the K-L expansion was to expand the random response process along a basis in the Hilbert space of random variables and to compute the coefficients in the expansion with the use of some error minimizing criterion. The Polynomial Chaoses defined in Chapter III were used as such an expansion and the Galerkin scheme was invoked to make the error orthogonal to the approximation space. The result of the procedure was a convergent expansion of the response process in terms of multidimensional orthogonal polynomials.

In Chapter IV the usefulness of the representation of the stochastic system response as derived in Chapter III was addressed from a perspective of determining the probabilistic characteristics of the response of the system. In this context elementary concepts from reliability theory were reviewed. The validity and limitations of the approximations involved were discussed. A method for reliability analysis was introduced that can be coupled with the system response representation presented in Chapter III. The method was based on viewing the Homogeneous Chaos expansion of the response process as providing an explicit expression of the limit state surface. Reliability related results could be readily obtained from this expression. One option in implementing the proposed method involved computing the moments of the response. These are obtainable, to any order, in a closed form. Alternatively, the cummulants of the response can be used. Another option was based on a multidimensional Edgeworth approximation for the probability distribution function in terms of multidimensional Hermite polynomials. The coefficients in this expansion could be obtained via a Galerkin scheme. Finally, a third option consisted of numerically simulating the Polynomial Chaoses resulting in a corresponding statistical population for the response process which was used in conjunction with nonparametric density estimation techniques to obtain approximations to the probability density function of the response.

The solution methods presented in Chapter III were applied in Chapter V to three problems of engineering mechanics. The first problem was that of a cantilever beam subjected to a static load. The modulus of elasticity of the beam was considered to be the realization of a homogeneous Gaussian process. The beam was discretized into ten finite elements and a variational formulation over the discrete mesh was used to obtain the associated algebraic random equations. Both solution procedures described in Chapter III were implemented to solve the resulting equations. The second problem involved a plate subjected to a static in-plane tension. Two cases were investigated. The first case involved a square plate; the second case involved a plate with curved boundary. Sixteen finite elements were used in both cases. Again a variational formulation was employed to obtain the governing discrete random equations. For both the beam problem and the plate problem, it was found that for the response process to achieve a certain accuracy, the terms needed in the final expression using the Homogeneous Chaos expansion were fewer than the terms required using the Neumann expansion approach. The method described in section (4.4) was employed to obtain approximations to the probability density function of the response of the beam and the plate. Good agreement was observed with the corresponding probability distributions that were deduced based on a large number of Monte Carlo simulations of the solution process. The third problem that was investigated involved a free-free beam lying on a linearly elastic foundation and subjected to a random excitation possessing both temporal and spatial random fluctuations. The modulus of reaction of the underlying foundation was also considered to be the realization of a homogeneous stochastic process. Following the same procedure as for the other examples, a discrete set of equations involving random variables was obtained. Again, the spectral approach was applied to derive an expression for the spectral density function of the response process. The analytical results were in good agreement with Monte Carlo simulation data except in the neighborhood of the natural frequency of the system where the Neumann expansion approximation seemed to diverge. This phenomenon was attributed to the fact that the matrices being inverted tended to be ill-conditioned at the vicinity of the resonance frequency and the requirements for the applicability of the Neumann expansion method were not satisfied. However, results from the Homogeneous Chaos approximation were well behaved and matched remarkably well the simulation data.

Some possible extensions of the concepts in this monograph are particularly notable, specifically in terms of improving the computational efficiency of the proposed method. As mentioned in Chapter V, the coefficients in the Homogeneous Chaos expansion obtained from lower expansions provide good initial guesses when used in the context of an iterative computational scheme. This fact suggests the possibility of implementing concepts from multigrid techniques (Hackbusch and Trottenberg, 1981) to increase the computational efficiency of the proposed methods. Another potential extension consists of implementing different basis functions for the Hilbert space of random variables. Specifically, rational functions in the uncorrelated random variables are an appealing choice. Of course, simulation methods will have to be applied in order to compute the coefficients c_{ijm} appearing in equation (3.72). However, these coefficients need to be calculated only once for a particular set of basis functions. The additional computational cost may be warranted by possible convergence acceleration.

It is believed that the material presented in this monograph constitutes a meaningful tool for accelerating the evolution of the field of Stochastic Finite Elements. Irrespective of the nature of future development in this field, such as static or dynamic problems, steady-state or transient solutions, and solid or fluid media, the pivotal concept of any analysis method will remain an expeditious and tractable representation of the randomness in the excitation, the medium, and the response; the spectral representation of random processes used herein is a viable option meriting further attention.

Bibliography

[1] Adomian, G., "Stochastic Green's functions", pp. 1-39, *Proceedings of Symposia in Applied Mathematics, Vol. 16: Stochastic Processes in Mathematical Physics and Engineering*, Edited by Richard Bellman, American Mathematical Society, Providence, Rhode Island, 1964.

[2] Adomian, G. and Malakian, K., "Inversion of stochastic partial differential operators - The linear case", *J. Math. Anal. Appl.*, Vol. 77, pp. 309-327, 1980.

[3] Adomian, G., *Stochastic Systems*, Academic Press, New York, 1983.

[4] Akin, J.E. *Application and Implementation of Finite Element Methods,*, Academic Press, New York, 1982.

[5] Aronszajn, N., "Theory of reproducing kernels", *Transactions of the American Mathematical Society*, Vol. 68, No. 3, pp. 337-404, May 1950.

[6] Babuska, I., Zienkiewicz, O.C., Gago, J. and de A. Oliveria, E.R., Editors, *Accuracy Estimates and Adaptive Refinements in Finite Element Computations*, John Wiley and Sons Ltd., New York, 1986.

[7] Beckers, Richard A. and Chambers, John M., *S: An Interactive Environment for Data Analysis and Graphics*, Wadsworth Statistics/Probability Series, Belmont, California, 1984.

[8] Benaroya, H., "Stochastic structural dynamics - a theoretical basis and a selective review of the literature", Weidlinger & Associates, Technical Report No.1, New York, April 1984.

[9] Benaroya, H. and Rehak, M., "Parametric excitations I: Exponentially correlated parameters", *Journal of Engineering Mechanics, ASCE*, Vol. 113, p. 861-874, June 1987.

[10] Benaroya, H. and Rehak, M., "Finite element methods in probabilistic structural analysis: A selective review", *Applied Mechanics Reviews*, Vol. 41, No. 5, pp. 201-213, May 1988.

[11] Benjamin, J. and Cornell, C., *Probability, Statistics and Decision for Civil Engineers*, McGraw-Hill, New York, 1970.

[12] Bharrucha-Reid, A.T., "On random operator equations in Banach space", *Bull. Acad. Polon. Sci., Ser. Sci. Math. Astr. Phys.*, Vol. 7, pp. 561-564, 1959.

[13] Bharrucha-Reid ed., *Probabilistic Methods in Applied Mathematics*, Academic Press, New York, 1968.

[14] Bose, A., *A Theory of Nonlinear Systems*, Massachussets Institute of Technology, RLE Report No. 309, July 1956.

[15] Boyce, E.W. and Goodwin, B.E., "Random transverse vibration of elastic beams", *SIAM Journal*, Vol. 12, No. 3, pp. 613-629, September, 1964.

[16] Brockett, R.W., "Volterra series and geometric control theory", *Automatica*, Vol. 12, pp. 167-176, 1976.

[17] Cakmak, A. and Sherif, R., "Parametric time series models for earthquake strong ground motions and their relationship to site parameters", *Proc. 8th World Conf. Earthq. Eng.*, San Francisco, pp. 581-588, 1984.

[18] Cameron, R.H. and Martin, W.T., "The orthogonal development of nonlinear functionals in series of Fourier-Hermite functionals", *Ann. Math*, Vol. 48, pp. 385-392, 1947.

[19] Chorin, A. J., "Hermite expansions in Monte-Carlo computation", *Journal of Computational Physics*, Vol. 8, pp. 472-482, 1971.

[20] Chorin, A. J., "Numerical study of slightly viscous flow", *Journal of Fluid Mechanics*, Vol. 57, Part 4, pp. 785-796, 1973.

[21] Collins, J.D. and Thompson, W.T., "The eigenvalue problem for structural systems with uncertain parameters", *AIAA Journal*, Vol.7, No.4, pp. 642-648, 1969.

[22] Cornell, C.A., "A probability-based structural code", *Journal of the American Concrete Institute*, Vol. 66, No. 12, December 1969.

[23] Courant and Hilbert, *Methods of Mathematical Physics*,, Interscience, New York, 1953.

[24] Cramer, H., "Stochastic processes as curves in Hilbert space", *Theory of Probability and its Applications*, Vol. IX, No. 2, pp. 169-179, 1964.

[25] Cruse, T.A., Wu, Y.T., Dias, J.B. and Rajagopal, K.R., "Probabilistic structural analysis methods and applications", *Computers and Structures*, Vol. 30, No. 1-2, pp. 163-170, 1988.

[26] Cruse, T.A. and Chamis, C.C., Editors, *Probabilistic Structural Analysis Methods (PSAM) for Select Space Propulsion Components*, Manuscripts and reprints by Southwest Research Institute, San Antonio, TX, May 1989.

[27] Dendrou, B. and Houstis, E., "An inference finite element model for field problems", *Appl. Math. Modeling*, Vol. 1, pp.109-114, 1978a.

[28] Dendrou, B. and Houstis, E., *Uncertainty Finite Element Analysis*, Technical Report CSD-TR 271, Purdue University, 1978b.

[29] Der-Kiureghian, A., and Liu, P.-L., "Structural reliability under incomplete probability information", *Journal of Engineering Mechanics, ASCE*, Vol. 112, No. 1, pp. 85-104, 1986.

[30] Der-Kiureghian, A., Lin, H.-Z. and Hwang, S.-J., "Second-order reliability approximations", *Journal of Engineering Mechanics, ASCE*, Vol. 113, No. 8, pp. 1208-1225, 1987a.

[31] Der-Kiureghian, A., "The stochastic finite element method in structural reliability", *Proceedings, U.S. Austria Joint Seminar on Stochastic Structural Mechanics*, Florida Atlantic university, Boca Raton, Florida, May 1987b.

[32] Devijver, P.A. and Kittler, J., *Pattern Recognition: A Statistical Approach*, Prentice-Hall, Englewood Cliffs, New Jersey, 1982.

[33] Ditlevsen, O., *Uncertainty Modelling*, McGraw-Hill, New York, 1981.

[34] Doob, J.L., *Stochastic Processes*, Wiley, New York, 1953.

[35] Engels, D.D., *The Multiple Stochastic Integral*, Memoirs of the American Mathematical Society, Vol. 38, No. 265, 1982.

[36] Fredholm, I., "Sur une class d'equations fonctionnelles", *Acta Mathematica*, Vol. 27, pp. 365-390, 1903.

[37] Freudenthal, A.M., "The safety of structures", *Transactions of the ASCE*, Vol. 112, pp. 125-180, 1947.

[38] Freudenthal, A.M., Garrelts, J.M., and Shinozuka, M., "The analysis of structural safety", *Journal of the Structural Engineering Division, ASCE*, Vol. 92, No. ST1, pp. 267-325, 1966.

[39] Gel'fand, I.M. and Vilenkin, N.Ya., *Generalized Functions, Vol. 4*, Academic Press, New York, 1964.

[40] George, P., *Continuous Nonlinear Systems*, Massachussets Institute of Technology, RLE Report No. 355, July 1959.

[41] Ghanem, R., Spanos, P. and Akin, E., "Orthogonal Expansion for Beam Variability", *Proceedings of the Conference on Probabilistic Engineering Mechanics, ASCE*, Blacksburg, Va., pp. 156-159, May 25-27, 1988.

[42] Ghanem, R.,, *Analysis of Stochastic Systems with Discrete Elements*, PhD Thesis, Rice University, Houston, TX, November 1988.

[43] Ghanem, R. and Spanos, P., "Galerkin Based Response Surface Approach for Reliability Analysis", *Proceedings of the 5th International Conference on Structural Safety and Reliability*, San Francisco, CA., pp. 1081-1088, August 8-11, 1989.

[44] Ghanem, R. and Spanos, P., "Polynomial Chaos in Stochastic Finite Elements", *Journal of Applied Mechanics, ASME*, Vol. 57, No.1, pp. 197-202, March 1990.

[45] Ghoniem, A.F. and Oppenheim, A.K., "Numerical solution for the problem of flame propagation by the random element method", *AIAA Journal*, Vol. 22, No. 10, pp. 1429-1435, October 1984.

[46] Golub, G. and Van Loan, C., *Matrix Computations*, Johns Hopkins University Press, Baltimore, 1984.

[47] Grad, H., "Note on N-dimensional hermite polynomials", *Communications in Pure and Applied Mathematics*, Vol. 2, pp. 325-330, December 1949.

[48] Grigoriu, M., Veneziano, D. and Cornell, C.A., "Probabilistic Modeling as Decision Making", *Journal of Engineering Mechanics, ASCE*, Vol. 105, EM4, pp. 585-676., August 1979.

[49] Grigoriu, M., "Methods for approximate reliability analysis", *Structural Safety*, Vol. 1, pp. 155-165, 1982.

[50] Hackbusch, W. and Trottenberg, U. Editors, *Multigrid Methods*, Proceedings of the Conference Held at Koln-Porz, November 23-27, Springer-Verlag, Berlin, 1981.

[51] Hans, O., "Random operator equations", *Proceedings of the Fourth Berkeley Symposium on Mathematical Statistics and Probability*, Neyman J. Editor, University of California, Vol. II, pp. 185-202, 1961.

[52] Harichandran, R. and Vanmarcke, E.H., "Stochastic variation of earthquake ground motion in space and time", *Journal of Engineering Mechanics, ASCE*, Vol. 112, No. EM2, pp. 154-174, 1986.

[53] Hart, G.C. and Collins, J.D., "The treatment of randomness in finite element modelling", *SAE Shock and Vibrations Symposium*, Los Angeles, CA, pp. 2509-2519, Oct. 1970.

[54] Hasofer, A.M. and Lind, N.C., "Exact and invariant second-moment code format", *Journal of Engineering Mechanics, ASCE*, Vol. 100, EM1, pp. 111-121, 1974.

[55] Hasselman, T.K. and Hart, G.C., "Modal analysis of random structural systems", *Journal of Engineering Mechanics, ASCE*, Vol. 98, No. EM3, 1972, pp. 561-579.

[56] Hida, T. and Ikeda, N., "Analysis on Hilbert space with reproducing kernel arising from multiple Wiener integral", *Proceedings of the Fifth Berkeley Symposium, part I*, Vol. 2, pp. 117-143, 1965.

[57] Hisada, T. and Nakagiri, S., "Role of the stochastic finite element method in structural safety and reliability", *ICOSSAR '85, International Conference on Structural Safety and Reliability*, pp. I-385, 1985.

[58] Hopf, E., "Statistical hydrodynamics and functional calculus", *J. Ratl. Mech. Anal.*, Vol. 1, pp. 87-123, 1952.

[59] Huang, S.T. and Cambanis, S., "Stochastic multiple Wiener integrals for gaussian processes", *The Annales of Probability*, Vol. 6, No. 4, pp. 585-614, 1978.

[60] Huang, S.T. and Cambanis, S., "Spherically invariant processes: Their nonlinear structure, discrimination, and estimation", *Journal of Multivariate Analysis*, Vol. 9, pp. 59-83, 1979.

[61] Ibrahim, R., "Structural dynamics with parameter uncertainties", *App. Mech. Rev.*, Vol. 40, No.3, pp. 309-328, 1987.

[62] Imamura, T., Meecham, W. and Siegel, A., "Symbolic calculus of the Wiener Process and Wiener-Hermite expansion", *Journal of Mathematical Physics*, Vol. 6, No. 5, pp.695-706, May 1965.

[63] Imamura, T. and Meecham, W., "Wiener-Hermite expansion in model turbulence in the late decay stage", *Journal of Mathematical Physics*, Vol. 6, No. 5, pp.707-721, May 1965.

[64] Itô, K., "Multiple Wiener integral", *Journal of the Mathematical Society of Japan*, Vol. 3, No. 1, pp. 157-169, May 1951.

[65] Iyengar, R. and Dash, P., "Random vibration analysis of stochastic time-varying systems", *Journal of Sound and Vibration*, Vol. 45, pp.69-89, 1976.

[66] Jahedi, A. and Ahmadi, G., "Application of Wiener-Hermite expansion to nonstationary random vibration of a Duffing oscillator", *Journal of Applied Mechanics, ASME*, Vol. 50, pp. 436-442, June 1983.

[67] Jordan, D.W. and Smith, P., *Nonlinear Ordinary Differential Equations*, Oxford University Press, Oxford, 1977.

[68] Juncosa, M., "An integral equation related to the Bessel functions", *Duke Mathematical Journal*, Vol. 12, pp. 465-468, 1945.

[69] Kac, M. and Siegert, A.J.F., "An explicit representation of a stationary gaussian process", *Ann. Math. Stat.*, Vol. 18, pp.438-442, 1947.

[70] Karhunen, K., "Uber lineare methoden in der wahrscheinlichkeitsrechnung", *Amer. Acad. Sci., Fennicade, Ser. A, I*, Vol. 37, pp. 3-79, 1947; (Translation: RAND Corporation, Santa Monica, California, Rep. T-131, Aug. 1960).

[71] Kailath, T., "Some integral equations with nonrational kernels", *IEEE, Transactions on Information Theory*, Vol. IT-12, No. 4, pp.442-447, October 1966

[72] Kailath, T., "A view of three decades of linear filtering theory", *IEEE, Transactions on Information Theory*, Vol. IT-20, pp. 146-181, 1974.

[73] Kakutani, S., "Spectral analysis of stationary gaussian processes", *Proceedings of the Fourth Berkeley Symposium on Mathematical Statistics and Probability*, Neyman J. Editor, University of California, Vol. II, pp. 239-247, 1961.

[74] Kallianpur, G., *Stochastic Filtering Theory*, Springer-Verlag, Berlin, 1980.

[75] Kolmogorov, A.N., "Sur l'interpolation et l'extrapolation des suites stationnaires", *Comptes Rendus de l'Academie des Sciences*, Vol. 208, pp. 2043, 1939.

[76] Kotulski, Z. and Sobczyk, K., "Characteristic functionals of randomly excited systems", *Physica*, Vol. 123A, pp. 261-278, 1984.

[77] Kozin, F., "A survey of stability of stochastic systems", *Automatica*, Vol. 5, pp. 95-112, 1969.

[78] Kree, P. and Soize, C., *Mathematics of Random Phenomena*, MIA, Reidel Publishing, Boston, Massachussets, 1986.

[79] Landau, H.J. and Pollak, H.O., "Prolate spheroidal wave functions; Fourier analysis and uncertainty - II", *Bell System Technical Journal*, pp. 65-84, 1961.

[80] Landau, H.J. and Pollak, H.O., "Prolate spheroidal wave functions; Fourier analysis and uncertainty - III; The dimension of the space of essentially time and band-limited signals", *Bell System Technical Journal*, pp. 1295-1336, 1962.

[81] Lawrence, M.A., "Basis random variables in finite element analysis", *Journal for Numerical Methods in Engineering*, Vol. 24, pp. 1849-1863, 1987.

[82] Lee, L.C., "Wave propagation in a random medium: a complete set of the moment equations with different wavenumbers", *J. Math. Phy.*, Vol. 15, No. 9, 1974.

[83] Li, C.W. and Blankenship, G. L., "Almost sure stability of linear stochastic systems with Poisson process coefficient", *SIAM, Journal on Applied Mathematics*, Vol. 46, No. 5, pp. 875-911, 1986.

[84] Lin, Y.K., *Probabilistic Theory of Structural Dynamics*, McGraw-Hill, New York, 1967.

[85] Lin, Y.K., Kozin, F., Wen, Y.K., Casciati, F., Schueller, G.I., Der Kiureghian, A., Ditlevsen, O., Vanmarcke, E.H., "Methods of stochastic structural dynamics", *Structural Safety*, Vol. 3, pp. 167-194, 1986.

[86] Liu, P.-L., Lin, H.-Z. and Der-Kiureghian, A., *CALREL User Manual.* Report No. UCB/SEMM-89/18, Department of Civil Engineering, University of California, Berkeley, CA, 1989.

[87] Liu, S.C., "Evolutionary power spectral density of strong-motion earthquakes", *Bulletin of the Seismological Society of America*, Vol. 60, No. 3, pp. 891-900, June 1970.

[88] Liu, W.K., Belytschko, T. and Mani, A., "Probabilistic finite elements for transient analysis in nonlinear continua", *Advances in Aerospace Structural Analysis, Proceedings ASME WAM*, Miami Beach, FL., Edited by O.H. Burnside and C.H. Pharr, Vol. AD-09, pp. 9-24, 1985.

[89] Liu, W.K., Besterfield, G. and Mani, A., "Probabilistic finite element methods in nonlinear structural dynamics", *Computer Methods in Applied Mechanics and Engineering*, Vol. 57, pp. 61-81, 1986.

[90] Liu, W.K., Besterfield, G. and Belytschko, T., "Transient probabilistic systems", *Computer Methods in Applied Mechanics and Engineering*, Vol. 67, pp. 27-54, 1988.

[91] Loeve, M., "Fonctions aleatoires du second ordre", supplement to P. Levy, *Processus Stochastic et Mouvement Brownien*, Paris, Gauthier Villars, 1948.

[92] Loeve, M., *Probability Theory, 4th edition*, Springer-Verlag, Berlin, 1977.

[93] *MACSYMA, Reference Manual, Version 12*, Symbolics Inc., 1986.

[94] Lumley, J.L., *Stochastic Tools in Turbulence*, Academic Press, New York, 1970.

[95] Madsen, H.O., Krenk, S. and Lind, N.C., *Methods of Structural Safety*, Prentice-Hall, Englewood Cliffs, New Jersey, 1986.

[96] Maltz, F.H. and Hitzl, D.L., "Variance reduction in Monte-Carlo computations using multi-dimensional Hermite polynomials", *Journal of Computational Physics*, Vol. 32, pp. 345-376, 1979.

[97] Masri, S. and Miller, R., "Compact probabilistic representation of random processes", *Journal of Applied Mechanics, ASME*, Vol. 49, pp. 871-876, 1982.

[98] Matheron, G., *Estimating and Choosing: An Essay on Probability in Practice*, Springer-Verlag, Berlin, 1989.

[99] McCullagh, P., "Tensor notation and cummulants of polynomials", *Biometrika*, Vol. 71, No. 3, pp. 461-476, 1984.

[100] McCullagh, P., *Tensor Methods in Statistics*, Chapman and Hall, London, 1987.

[101] McKean, H.P., "Geometry of differential space", *The Annales of Probability*, Vol. 1, No. 2, 1973, pp.197-206.

[102] Melchers, R., *Structural Reliability, Analysis and Prediction*, Ellis Horwood Limited, W. Sussex, England, 1987.

[103] Mignolet, M., *ARMA Simulation of Multivariate and Multidimensional Random Processes*, Thesis presented in partial fulfillment of the degree of Ph.D, Rice University, Houston, TX., 1987.

[104] Mikhlin, S.G., *Integral Equations*, Pergamon Press, Oxford, 1957.

[105] Millwater, H., Palmer and Fink, P., "NESSUS/EXPERT - An expert system for probabilistic structural analysis methods", *29th AIAA/ASME/ASCE/AHS/ASC Structures, Structural Dynamics, and Materials Conference*, Williamsburg, Virginia, April, 18-20, 1988.

[106] Millwater, H., Wu, Y.-T., Dias, J.B., McClung, R.C., Raveendra, S.T. and Thacker, B.H., "The NESSUS software system for probabilistic structural analysis", *ICOSSAR '89, International Conference on Structural Safety and Reliability*, San Francisco, California, August 8-11, 1989.

[107] Murray, F.J. and Von Neumann, J., "On rings of operators", *Annales of Mathematics*, Vol. 37, pp. 116-229, 1936.

[108] Nakagiri, S. and Hisada, T., "Stochastic finite element method applied to structural analysis with uncertain parameters", *Proc. Intl. Conference on FEM*, pp. 206-211, August 1982.

[109] Nashed, Z., Editor, *Generalized Inverses and Applications, Proceedings of an Advanced Seminar*, The University of Wisconsin-Madison, October 8-10, 1973, Academic Press, New York, 1976.

[110] Nayfeh, A.H., *Perturbation Methods*, John Wiley and Sons, New York, 1973.

[111] Noble, B., *The Wiener-Hopf Technique*, Pergamon Press, London, 1958.

[112] Oden J. T., *Applied Functional Analysis*, Prentice-Hall, Englewood Cliffs, New Jersey, 1979.

[113] Ogura, H., "Orthogonal functionals of the Poisson process", *IEEE, Transactions on Information Theory*, Vol. IT-18, No. 4, pp.473-481, July 1972.

[114] Parzen, E., *Statistical Inference on Time Series by Hilbert Space Methods*, Technical Report No. 23, O.N.R. Contract 225(21), Statistics Department, Stanford University, CA, January 1959.

[115] Paley, R. and Wiener, N., *Fourier Transforms in the Complex Domain*, American Mathematical Society Colloquium Publications Volume XIX, New York, 1934.

[116] Palm, G. and Poggio, T., "The Volterra representation and the Wiener expansion: validity and pitfalls", *SIAM, Journal of Applied Mathematics*, Vol. 33, No. 2, September 1977, pp. 195-216.

[117] Polhemus, N.W. and Cakmak, A.S., "Simulation of earthquake ground motion using ARMA models", *Earthquake Engineering and Structural Dynamics*, Vol. 9, pp. 343-354, 1981.

[118] Priestley, M.B., *Non-Linear and Non-Stationary Time Series Analysis*, Academic Press, New York, 1988.

[119] Rabiner, L.R. and Gold, B., *Theory and Applications of Digital Signal Processing*, Englewood Cliffs, New Jersey, 1975.

[120] Rao, T.S. and Gabr, M.M., *An Introduction to Bispectral Analysis and Bilinear Times Series Models*, Springer-Verlag, Berlin, 1984.

[121] Rektorys, K., *Variational Methods in Mathematics, Science and Engineering*, 2nd edition, D. Reidel Publishing Company, Boston, Massachussets, 1980.

[122] Roberts, S., "Convergence of a random walk method for the Burges equation", *Mathematics of Computation*, Vol. 52, No. 186, pp. 647-673, 1989.

[123] Rosenblatt, M., "Remarks on a multivariate transformation", *Ann. Math. Stat.*, Vol. 23, pp. 470-472, 1952.

[124] Rugh, W., *Nonlinear System Theory*, The John Hopkins University Press, Baltimore and London, 1981.

[125] Schetzen, M., *The Volterra and Wiener Theories and Nonlinear Systems*, John Wiley and Sons Inc., New York, 1980.

[126] Schueller, G. and Stix, R., "A critical appraisal of methods to determine failure probabilities", *Structural Safety*, Vol. 4, pp. 293-309, 1987.

[127] Segall, A. and Kailath, T., "Orthogonal functionals of independent-increment processes", *IEEE, Transactions on Information Theory*, Vol. IT-22, No. 3, pp. 287-298, May 1976.

[128] Shames, I.H, and Dym, C.L., *Energy and Finite Element Methods in Structural Mechanics*, Hemisphere Publishing Corporation, New York, 1985.

[129] Shinozuka, M. and Astill, J., "Random eigenvalue problems in structural mechanics", *AIAA Journal*, Vol. 10, No. 4, pp. 456-462, 1972.

[130] Shinozuka, M. and Jan, C.M., "Digital simulation of random Processes and its applications", *Journal of Sound and Vibration*, Vol. 25, No. 1, 1972.

[131] Shinozuka, M., "Digital simulation of random processes in engineering mechanics with the aid of FFT techniques", *Stochastic Problems in Mechanics*, Edited by S.T. Ariaratnan and H.H.E. Leipholtz, University of Waterloo Press, Waterloo, 1974.

[132] Shinozuka, M. and Lenoe, E., "A probabilistic model for spatial distribution of material properties", *Eng. Fracture Mechanics*, Vol. 8, pp. 217-227, 1976.

[133] Shinozuka, M. and Nomoto, T., *Response Variability Due to Spatial Randomness of Material Properties*, Technical Report, Dept. of Civil Engineering, Columbia University, New York, 1980.

[134] Shinozuka, M., "Basic Analysis for Structural Safety", *Journal of Structural Engineering, ASCE*, Vol. 109, No. 3, pp. 721-740, March 1983.

[135] Shinozuka, M., "Stochastic fields and their digital simulation", *Lecture Notes for the CISM Course on Stochastic Methods in Structural Mechanics*, Udine, Italy, 1985.

[136] Shinozuka, M. and Deodatis, G., *Response Variability of Stochastic Finite Element Systems*, Technical Report, Dept. of Civil Engineering, Columbia University, New York, 1986.

[137] Shinozuka, M., "Structural response variability", *Journal of Engineering Mechanics, ASCE*, Vol. 113, No. EM6, pp. 825-842, 1987.

[138] Shinozuka, M. (Editor), *Stochastic Mechanics, Vol. I*, Department of Civil Engineering, Columbia University, New York, 1987.

[139] Slepian, D. and Pollak, H.O., "Prolate spheroidal wave functions; Fourier analysis and uncertainty - I", *Bell System Technical Journal*, pp. 43-63, 1961.

[140] Slepian, D., "Prolate spheroidal wave functions; Fourier analysis and uncertainty - IV: Extensions to many dimensions; Generalized prolate spheroidal functions", *Bell System Technical Journal*, pp. 3009-3057, 1964.

[141] Slepian, D., "Some asymptotic expansions for prolate spheroidal wave functions", *Journal of Mathematics and Physics*, Vol. 44, No. 2, pp. 99-140, 1965.

[142] Slepian, D., "A numerical method for determining the eigenvalues and eigenfunctions of analytic kernels", *SIAM, Journal of Numerical Analysis*, Vol. 5, No. 3, pp. 586-600, September 1968.

[143] Slepian, D., "Prolate spheroidal wave functions; Fourier analysis and uncertainty - V", *Bell System Technical Journal*, pp. 1371-1430, 1977.

[144] Sobczyk, K., *Wave Propagation in Random Media*, Elsevier, Amsterdam, 1985.

[145] Soong, T. and Bogdanoff, J., "On the natural frequencies of a disordered linear chain of n degrees of freedom", *Int. J. Mech. Sci.*, Vol. 5, pp. 237-265, 1963.

[146] Spanos, P.D. and Hansen, J., "Linear prediction theory for digital simulation of sea waves", *Journal of Energy Resources Technology*, Vol. 103, pp. 243-249, 1981.

[147] Spanos, P.D. and Mignolet, M., "Z-Transform modeling of P-M wave spectrum", *Journal of Engineering Mechanics, ASCE*, Vol. 112, No. EM 8, pp. 745-756, 1986.

[148] Spanos P.D. and Ghanem R., "Stochastic finite element expansion for random media", *Journal of the Engineering Mechanics Division, ASCE*, Vol. 115, No. 5, pp. 1035-1053, May 1989.

[149] Thoft-Christensen, P. and Baker, M., *Structural Reliability and Its Applications*, Springer-Verlag, Berlin, 1982.

[150] Thoft-Christensen, P., Editor, *Reliability Theory and its Application in Structural and Soil Mechanics*, NATO ASI Series, Martinus Nijhoff Publishers, The Hague, Denmark, 1983.

[151] Traina, M.I., Miller, R.K. and Masri, S.F., "Orthogonal decomposition and transmission of nonstationary random processes", *Probabilistic Engineering Mechanics*, Vol.1, No.3, pp. 136-149, September 1986.

[152] Vanmarcke, E.H., "Stochastic finite element analysis", pp. 278-294, *Probabilistic Methods in Structural Engineering*, Edited by Shinozuka, M. and Yao, J.T.Y., ASCE, 1981.

[153] Vanmarcke, E.H., *Random Fields*, MIT Press, Cambridge Mass., 1983.

[154] Vanmarcke, E.H. and Grigoriu, M., "Stochastic finite element analysis of simple beams", *Journal of the Engineering Mechanics Division, ASCE*, Vol. 109, No. EM5, pp. 1203-1214, 1983.

[155] Van Trees, H.L., *Detection, Estimation and Modulation Theory, Part 1*, Wiley, New York 1968.

[156] Volterra, V., *Lecons sur les Equations Integrales et Integrodifferentielles*, Paris: Gauthier Villars, 1913.

[157] Wiener, N., "Differential space", *J. Math. Phys.*, Vol. 2, pp. 131-174, 1923.

[158] Wiener, N., "The homogeneous chaos", *Amer. J. Math*, Vol. 60, pp. 897-936, 1938.

[159] Wiener, N., *Extrapolation, Interpolation and Smoothing of Stationary Time Series with Engineering Applications*, Technology Press and Wiley, New York, 1949.

[160] Wiener, N., *Nonlinear Problems in Random Theory*, Technology Press of the Massachussets Institute of Technology and John Wiley and Sons Inc., New York, 1958.

[161] Wintner, A. and Wiener, N., "The discrete chaos", *American Journal of Mathematics*, Vol. 65, pp. 279-298, 1943.

[162] Wirshing, P.H. and Wu, Y.-T., "Advanced reliability methods for structural evaluation", *Journal of Engineering for Industry, ASME*, Vol. 109, pp. 19-23, February 1987.

[163] Wong, E. and Zakai, M., "On the relation between ordinary and stochastic differential equations", *Int. J. Engng. Sci.*, Vol. 3, pp. 213-228, 1965.

[164] Wu, Y.-T. and Wirshing, P.H., "New algorithm for structural reliability estimation", *Journal of Engineering Mechanics, ASCE*, Vol. 113, No. 9, pp. 1319-1336, September, 1987.

[165] Wu, Y.-T., "Demonstration of a new, fast probability integration method for reliability analysis", *Journal of Enginering for Industry, ASME*, Vol. 109, pp. 24-28, February 1987.

[166] Yaglom, A.M., *An Introduction to the Theory of Stationary Random Functions*, Prentice-Hall, Englewood Cliffs, New Jersey, 1962.

[167] Yamazaki, F., Shinozuka, M. and Dasgupta, G., *Neumann Expansion for Stochastic Finite Element Analysis*, Technical Report, Department of Civil Engineering, Columbia University, New York, December 1985.

[168] Yamazaki, F. and Shinozuka, M., *Digital Generation of Non-Gaussian Stochastic Fields*, Technical Report, Department of Civil Engineering, Columbia University, New York, May 1986.

[169] Yamazaki, F., *Simulation of Stochastic Fields and its Applications to Finite Element Analysis*, ORI Report 87-04, Ohzaki Research Institute Inc., November 1987.

[170] Yang, T.Y. and Kapania, R.K., "Finite element random response analysis of cooling tower", *Journal of Engineering Mechanics, ASCE*, Vol. 110, No. EM4, pp. 589-609, 1984.

[171] Yasui, S., "Stochastic functional Fourier series, Volterra series and nonlinear systems analysis", *IEEE, Transactions on Automatic Control*, Vol. AC-24, No. 2, pp. 230-242, April 1979.

[172] Youla, D.C., "The solution of a homogeneous Wiener-Hopf integral equation occurring in the expansion of second-order stationary random functions", *IRE, Transactions on Information Theory*, pp. 18-193, September 1957.

[173] Zienkiewicz, O.C. and Taylor, R.L., *The Finite Element Method, Fourth Edition*, McGraw-Hill, New York, 1989.

Additional References

[01] Adhikari, S. and Manohar, C.S., "Transient dynamics of stochastically parametered beams," *J Eng Mech—ASCE,* Vol. 126, No. 11, pp. 1131–1140, Nov. 2000.

[02] Anders, M. and Hori, M., "Three-dimensional stochastic finite element method for elasto-plastic bodies," *International Journal for Numerical Methods in Engineering,* Vol. 51, No. 4, pp. 449–478, 2001.

[03] Bhattacharyya, B. and Chakraborty, S., "Sensitivity statistics of 3D structures under parametric uncertainty," *Journal of Engineering Mechanics—ASCE,* Vol. 127, No. 9, pp. 909–914, 2001.

[04] Brzakala, W. and Elishakoff, I., "Lessons pertaining to the finite element method for stochastic problems, learned from simplest example," *Chaos Solitons & Fractals,* Vol. 12, No. 7, pp. 1217–1232, 2001.

[05] Chakraborty, S. and Dey, S.S., "Stochastic finite element simulation of random structure on uncertain foundation under random loading," *International Journal of Mechanical Sciences,* Vol. 38, No. 11, pp. 1209–1218, 1996.

[06] Clouteau, D., Savin, E., and Aubry, D., "Stochastic simulations in dynamic soil-structure interaction," *Meccanica,* Vol. 36, No. 4, pp. 379–399, 2001.

[07] Deb, M.K., Babuska, I.M., and Oden, J.T., "Solution of stochastic partial differential equations using Galerkin finite element techniques," *Computer Methods in Applied Mechanics and Engineering,* Vol. 190, No. 48, pp. 6359–6372, 2001.

[08] Der Kiureghian, A. and Zhang, Y., "Space-variant finite element reliability analysis," *Computer Methods in Applied Mechanics and Engineering,* Vol. 168, Nos. 1–4, pp. 173–183, 1999.

211

[09]   Ditlevsen, O. and Tarp-Johansen, N.J., "Choice of input fields in stochastic finite elements," *Probabilistic Engineering Mechanics,* Vol. 14, Nos. 1–2, pp. 63–72, 1999.

[10]   Elishakoff, I. and Ren, Y.J., "The bird's eye view on finite element method for structures with large stochastic variations," *Computer Methods in Applied Mechanics and Engineering,* Vol. 168, Nos. 1–4, pp. 51–61, 1999.

[11]   Galal, O.H., El-Tawil, M.A., and Mahmoud, A.A., "Stochastic beam equations under random dynamic loads," *International Journal of Solids and Structures,* Vol. 39, No. 4, pp. 1031–1040, 2002.

[12]   Ghanem, R., "Higher-order sensitivity of heat conduction problems to random data using the spectral stochastic finite element method," *Journal of Heat Transfer—Transactions of the ASME,* Vol. 121, No. 2, pp. 290–299, 1999.

[13]   Ghanem, R., "Hybrid stochastic finite elements and generalized Monte Carlo simulation," *Journal of Applied Mechanics—Transactions of the ASME,* Vol. 65, No. 4, pp. 1004–1009, 1998.

[14]   Ghanem, R., "Ingredients for a general purpose stochastic finite elements implementation," *Computer Methods in Applied Mechanics and Engineering,* Vol. 168, Nos. 1–4, pp. 19–34, 1999.

[15]   Ghanem, R., "Probabilistic characterization of transport in heterogeneous media," *Computer Methods in Applied Mechanics and Engineering,* Vol. 158, Nos. 3–4, pp. 199–220, 1998.

[16]   Ghanem, R., "Scales of fluctuation and the propagation of uncertainty in random porous media," *Water Resources Research,* Vol. 34, No. 9, pp. 2123–2136, 1998.

[17]   Ghanem, R., "Stochastic finite elements with multiple random non-Gaussian properties," *Journal of Engineering Mechanics—ASCE,* Vol. 125, No. 1, pp. 26–40, 1999.

[18]   Ghanem, R., "The nonlinear Gaussian spectrum of log-normal stochastic processes and variables," *Journal of Applied Mechanics—Transactions of the ASME,* Vol. 66, No. 4, pp. 964–973, 1999.

[19]   Ghanem, R. and Brzakala, W., "Stochastic finite-element analysis of soil layers with random interface," *Journal of Engineering Mechanics—ASCE,* Vol. 122, No. 4, pp. 361–369, 1996.

[20]   Ghanem, R. and Dham, S., "Stochastic finite element analysis for multiphase flow in heterogeneous porous media," *Transport in Porous Media,* Vol. 32, No. 3, pp. 239–262, 1998.

[21]   Ghanem, R.G. and Kruger, R.M., "Numerical solution of spectral stochastic finite element systems," *Computer Methods in Applied Mechanics and Engineering,* Vol. 129, No. 3, pp. 289–303, 1996.

[22]   Ghanem, R. and Pellissetti, M., "Adaptive data refinement in the spectral stochastic finite element method," *Communications in Numerical Methods in Engineering,* Vol. 18, No. 2, pp. 141–151, 2002.

[23]   Ghanem, R. and Red-Horse, J., "Propagation of probabilistic uncertainty in complex physical systems using a stochastic finite element approach," *Physica D,* Vol. 133, Nos. 1–4, pp. 137–144, 1999.

[24]   Ghiocel, D.M. and Ghanem, R.G., "Stochastic finite-element analysis of seismic soil-structure interaction," *Journal of Engineering Mechanics—ASCE,* Vol. 128, No. 1, pp. 66–77, 2002.

[25]   Gutierrez, E. and Zaldivar, J.M., "The application of Karhunen-Loeve, or principal component analysis method, to study the nonlinear seismic response of structures," *Earthquake Engineering & Structural Dynamics,* Vol. 29, No. 9, pp. 1261–1286, 2000.

[26]   Holmes, J.D., Sankaran, R., Kwok, K.C.S., and Syme, M.J., "Eigenvector modes of fluctuating pressures on low-rise building models," *Journal of Wind Engineering and Industrial Aerodynamics,* Vol. 71, pp. 697–707, 1997.

[27]   Hurtado, J.E., "Analysis of one-dimensional stochastic finite elements using neural networks," *Probabilistic Engineering Mechanics,* Vol. 17, No. 1, pp. 35–44, 2002.

[28]   Hurtado, J.E. and Barbat. A.H., "Monte Carlo techniques in computational stochastic mechanics," *Archives of Computational Methods in Engineering,* Vol. 5, No. 1, pp. 3–29, 1998.

[29] Isukapalli, S.S., Roy, A., and Georgopoulos, P.G., "Efficient sensi-
      tivity/uncertainty analysis using the combined stochastic response
      surface method and automated differentiation: Application to envi-
      ronmental and biological systems," *Risk Analysis*, Vol. 20, No. 5, pp.
      591–602, Oct. 2000.

[30] Isukapalli, S.S., Roy, A., and Georgopoulos, P.G., "Stochastic
      Response Surface Methods (SRSMs) for uncertainty propagation:
      Application to environmental and biological systems," *Risk Analysis*,
      Vol. 18, No. 3, pp. 351–363, June 1998.

[31] Iwan, W.D. and Huang, C.T., "On the dynamic response of non-lin-
      ear systems with parameter uncertainties," *International Journal of
      Non-Linear Mechanics*, Vol. 31, No. 5, pp. 631–645, 1996.

[32] Iwan, W.D. and Jensen, H., "On the dynamic-response of continu-
      ous systems including model uncertainty," *Journal of Applied
      Mechanics—Transactions of the ASME*, Vol. 60, No. 2, pp. 484–490,
      1993.

[33] Jensen, H. and Iwan, W.D., "Response of systems with uncertain
      parameters to stochastic excitation," *Journal of Engineering
      Mechanics—ASCE*, Vol. 118, No. 5, pp. 1012–1025, 1992.

[34] Jensen, H. and Iwan, W.D., "Response variability in structural
      dynamics," *Earthquake Engineering & Structural Dynamics*, Vol. 20,
      No. 10, pp. 949–959, 1991.

[35] Karniadakis, G., "Quantifying uncertainty in CFD," *Journal of
      Fluids Engineering—Transactions of the ASME*, Vol. 124, No. 1, pp.
      2–3, Mar. 2002.

[36] Katafygiotis, L.S. and Beck, J.L., "A very efficient moment calcula-
      tion method for uncertain linear dynamic-systems," *Probabilistic
      Engineering Mechanics*, Vol. 10, No. 2, pp. 117–128, 1995.

[37] Katafygiotis, L.S. and Papadimitriou, C., "Dynamic response vari-
      ability of structures with uncertain properties," *Earthquake
      Engineering & Structural Dynamics*, Vol. 25, No. 8, pp. 775–793,
      1996.

[38] Le Maitre, O.P., Knio, O.M., Najm, H.N., and Ghanem, R.G., "A stochastic projection method for fluid flow I. Basic formulation," *Journal of Computational Physics,* Vol. 173, No. 2, pp. 481–511, 2001.

[39] Li, C.C. and Der Kiureghian, A., "Optimal discretization of random-fields," *Journal of Engineering Mechanics—ASCE,* Vol. 119, No. 6, pp. 1136–1154, 1993.

[40] Li, J. and Liao, S.T., "Response analysis of stochastic parameter structures under non-stationary random excitation," *Computational Mechanics,* Vol. 27, No. 1, pp. 61–68, 2001.

[41] Li, R. and Ghanem, R., "Adaptive polynomial chaos expansions applied to statistics of extremes in nonlinear random vibration," *Probabilistic Engineering Mechanics,* Vol. 13, No. 2, pp. 125–136, 1998.

[42] Manolis, G.D., "Green-function for wave motion in acoustic media with large randomness," *Journal of Sound and Vibration,* Vol. 179, No. 3, pp. 529–545, 1995.

[43] Manolis, G.D., "Stochastic soil dynamics," *Soil Dynamics and Earthquake Engineering,* Vol. 22, No. 1, pp. 3–15, 2002.

[44] Manolis, G.D. and Pavlou, S., "Fundamental solutions for SH-waves in a continuum with large randomness," *Engineering Analysis with Boundary Elements,* Vol. 23, No. 9, pp. 721–736, 1999.

[45] Matthies, H.G., Brenner, C.E., Bucher, C.G., and Soares, C.G., "Uncertainties in probabilistic numerical analysis of structures and solids—Stochastic finite elements," *Structural Safety,* Vol. 19, No. 3, pp. 283–336, 1997.

[46] Matthies, H.G. and Bucher, C., "Finite elements for stochastic media problems," *Computer Methods in Applied Mechanics and Engineering,* Vol. 168, Nos. 1–4, pp. 3–17, 1999.

[47] Matthies, H. and Keese, A., "Multilevel methods for stochastic systems," *Proceedings of the 2nd European Conference on Computational Mechanics,* June 26–29, 2001, Cracow, Poland.

[48] Mei, H., Agrawal, O.P. and Pai, S.S., "Wavelet-based model for stochastic analysis of beam structures," *AIAA Journal,* Vol. 36, No. 3, pp. 465–470, 1998.

[49] Nair, P.B. and Keane, A.J., "Stochastic reduced basis methods," *AIAA Journal,* Vol. 40, No. 8, pp. 1653–1664, 2002.

[50] Pan, W.W., Tatang, M.A., McRae, G.J., and Prinn, R.G., "Uncertainty analysis of direct radiative forcing by anthropogenic sulfate aerosols," *Journal of Geophysical Research—Atmospheres,* Vol. 102, No. D18, pp. 21915–21924, Sept. 27, 1997.

[51] Papadimitriou, C., Katafygiotis, L.S., and Beck, J.L., "Approximate analysis of response variability of uncertain linear systems," *Probabilistic Engineering Mechanics,* Vol. 10, No. 4, pp. 251–264, 1995.

[52] Pellissetti, M.F. and Ghanem, R.G., "Iterative solution of systems of linear equations arising in the context of stochastic finite elements," *Advances in Engineering Software,* Vol. 31, Nos. 8–9, pp. 607–616, 2000.

[53] Sakamoto, S. and Ghanem, R., "Polynomial chaos decomposition for the simulation of non-Gaussian nonstationary stochastic processes," *Journal of Engineering Mechanics—ASCE,* Vol. 128, No. 2, pp. 190–201, 2002.

[54] Savin, E. and Clouteau, D., "Elastic wave propagation in a 3-D unbounded random heterogeneous medium coupled with a bounded medium. Application to seismic soil-structure interaction (SSSI)," *International Journal for Numerical Methods in Engineering,* Vol. 54, No. 4, pp. 607–630, 2002.

[55] Schueller, G.I., "Computational stochastic mechanics—recent advances," *Computers & Structures,* Vol. 79, Nos. 22–25, pp. 2225–2234, 2001.

[56] Schueller, G.I., "Special issue—A state-of-the-art report on computational stochastic mechanics—Preface," *Probabilistic Engineering Mechanics,* Vol. 12, No. 4, pp. 197–313, 1997.

[57] Solari, G. and Carassale, L., "Modal transformation tools in structural dynamics and wind engineering," *Wind and Structures,* Vol. 3, No. 4, pp. 221–241, 2000.

[58] Spanos, P.D. and Ghanem, R., "Boundary element formulation for random vibration problems," *Journal of Engineering Mechanics— ASCE,* Vol. 117, No. 2, pp. 409–423, 1991.

[59] Spanos, P.D. and Zeldin, B.A., "Galerkin sampling method for stochastic mechanics problems," *Journal of Engineering Mechanics— ASCE,* Vol. 120, No. 5, pp. 1091–1106, 1994.

[60] Venini, P. and Mariani, C., "Free vibrations of uncertain composite plates via stochastic Rayleigh-Ritz approach," *Computers & Structures,* Vol. 64, Nos. 1–4, pp. 407–423, 1997.

[61] Wang, C., Tatang, M.A., and McRae, G.J., "Uncertainty analysis of reaction models based on the deterministic equivalent modeling method," *Abstracts of Papers of the American Chemical Society,* Vol. 216, No. 063-CINF, Part 1, Aug. 1998.

[62] Xiu, D.B., Lucor, D., Su, C.H., and Karniadakis, G., "Stochastic modeling of flow-structure interactions using generalized polynomial chaos," *Journal of Fluids Engineering—Transactions of the ASME,* Vol. 124, No. 1, pp. 51–59, Mar. 2002.

[63] Yeh, C.H. and Rahman, M.S., "Stochastic finite element methods for the seismic response of soils," *International Journal for Numerical and Analytical Methods in Geomechanics,* Vol. 22, No. 10, pp. 819–850, 1998.

[64] Zhang, J. and Ellingwood, B., "Effects of uncertain material properties on structural stability," *Journal of Structural Engineering— ASCE,* Vol. 121, No. 4, pp. 705–716, 1995.

[65] Zhang, J. and Ellingwood, B., "Orthogonal series expansions of random-fields in reliability-analysis," *Journal of Engineering Mechanics—ASCE,* Vol. 120, No. 12, pp. 2660–2677, 1994.

Index

A CATALOG OF SELECTED
DOVER BOOKS
IN SCIENCE AND MATHEMATICS

A CATALOG OF SELECTED

DOVER BOOKS

IN SCIENCE AND MATHEMATICS

Astronomy

BURNHAM'S CELESTIAL HANDBOOK, Robert Burnham, Jr. Thorough guide to the stars beyond our solar system. Exhaustive treatment. Alphabetical by constellation: Andromeda to Cetus in Vol. 1; Chamaeleon to Orion in Vol. 2; and Pavo to Vulpecula in Vol. 3. Hundreds of illustrations. Index in Vol. 3. 2,000pp. 6⅛ x 9¼.
23567-X, 23568-8, 23673-0 Three-vol. set

THE EXTRATERRESTRIAL LIFE DEBATE, 1750–1900, Michael J. Crowe. First detailed, scholarly study in English of the many ideas that developed from 1750 to 1900 regarding the existence of intelligent extraterrestrial life. Examines ideas of Kant, Herschel, Voltaire, Percival Lowell, many other scientists and thinkers. 16 illustrations. 704pp. 5⅜ x 8½. 40675-X

A HISTORY OF ASTRONOMY, A. Pannekoek. Well-balanced, carefully reasoned study covers such topics as Ptolemaic theory, work of Copernicus, Kepler, Newton, Eddington's work on stars, much more. Illustrated. References. 521pp. 5⅜ x 8½.
65994-1

AMATEUR ASTRONOMER'S HANDBOOK, J. B. Sidgwick. Timeless, comprehensive coverage of telescopes, mirrors, lenses, mountings, telescope drives, micrometers, spectroscopes, more. 189 illustrations. 576pp. 5⅝ x 8¼. (Available in U.S. only.)
24034-7

STARS AND RELATIVITY, Ya. B. Zel'dovich and I. D. Novikov. Vol. 1 of *Relativistic Astrophysics* by famed Russian scientists. General relativity, properties of matter under astrophysical conditions, stars, and stellar systems. Deep physical insights, clear presentation. 1971 edition. References. 544pp. 5⅝ x 8¼. 69424-0

Chemistry

CHEMICAL MAGIC, Leonard A. Ford. Second Edition, Revised by E. Winston Grundmeier. Over 100 unusual stunts demonstrating cold fire, dust explosions, much more. Text explains scientific principles and stresses safety precautions. 128pp. 5⅜ x 8½. 67628-5

THE DEVELOPMENT OF MODERN CHEMISTRY, Aaron J. Ihde. Authoritative history of chemistry from ancient Greek theory to 20th-century innovation. Covers major chemists and their discoveries. 209 illustrations. 14 tables. Bibliographies. Indices. Appendices. 851pp. 5⅜ x 8½. 64235-6

CATALYSIS IN CHEMISTRY AND ENZYMOLOGY, William P. Jencks. Exceptionally clear coverage of mechanisms for catalysis, forces in aqueous solution, carbonyl- and acyl-group reactions, practical kinetics, more. 864pp. 5⅜ x 8½.
65460-5

THE HISTORICAL BACKGROUND OF CHEMISTRY, Henry M. Leicester. Evolution of ideas, not individual biography. Concentrates on formulation of a coherent set of chemical laws. 260pp. 5⅜ x 8½. 61053-5

A SHORT HISTORY OF CHEMISTRY, J. R. Partington. Classic exposition explores origins of chemistry, alchemy, early medical chemistry, nature of atmosphere, theory of valency, laws and structure of atomic theory, much more. 428pp. 5⅜ x 8½. (Available in U.S. only.) 65977-1

GENERAL CHEMISTRY, Linus Pauling. Revised 3rd edition of classic first-year text by Nobel laureate. Atomic and molecular structure, quantum mechanics, statistical mechanics, thermodynamics correlated with descriptive chemistry. Problems. 992pp. 5⅜ x 8½. 65622-5

Engineering

DE RE METALLICA, Georgius Agricola. The famous Hoover translation of greatest treatise on technological chemistry, engineering, geology, mining of early modern times (1556). All 289 original woodcuts. 638pp. 6¾ x 11. 60006-8

FUNDAMENTALS OF ASTRODYNAMICS, Roger Bate et al. Modern approach developed by U.S. Air Force Academy. Designed as a first course. Problems, exercises. Numerous illustrations. 455pp. 5⅜ x 8½. 60061-0

DYNAMICS OF FLUIDS IN POROUS MEDIA, Jacob Bear. For advanced students of ground water hydrology, soil mechanics and physics, drainage and irrigation engineering and more. 335 illustrations. Exercises, with answers. 784pp. 6⅛ x 9¼. 65675-6

ANALYTICAL MECHANICS OF GEARS, Earle Buckingham. Indispensable reference for modern gear manufacture covers conjugate gear-tooth action, gear-tooth profiles of various gears, many other topics. 263 figures. 102 tables. 546pp. 5⅜ x 8½. 65712-4

MECHANICS, J. P. Den Hartog. A classic introductory text or refresher. Hundreds of applications and design problems illuminate fundamentals of trusses, loaded beams and cables, etc. 334 answered problems. 462pp. 5⅜ x 8½. 60754-2

MECHANICAL VIBRATIONS, J. P. Den Hartog. Classic textbook offers lucid explanations and illustrative models, applying theories of vibrations to a variety of practical industrial engineering problems. Numerous figures. 233 problems, solutions. Appendix. Index. Preface. 436pp. 5⅜ x 8½. 64785-4

STRENGTH OF MATERIALS, J. P. Den Hartog. Full, clear treatment of basic material (tension, torsion, bending, etc.) plus advanced material on engineering methods, applications. 350 answered problems. 323pp. 5⅜ x 8½. 60755-0

A HISTORY OF MECHANICS, René Dugas. Monumental study of mechanical principles from antiquity to quantum mechanics. Contributions of ancient Greeks, Galileo, Leonardo, Kepler, Lagrange, many others. 671pp. 5⅜ x 8½. 65632-2

METAL FATIGUE, N. E. Frost, K. J. Marsh, and L. P. Pook. Definitive, clearly writ-ten, and well-illustrated volume addresses all aspects of the subject, from the histori-cal development of understanding metal fatigue to vital concepts of the cyclic stress that causes a crack to grow. Includes 7 appendixes. 544pp. 5⅜ x 8½. 40927-9

STATISTICAL MECHANICS: Principles and Applications, Terrell L. Hill. Standard text covers fundamentals of statistical mechanics, applications to fluctuation theory, imperfect gases, distribution functions, more. 448pp. 5⅜ x 8½. 65390-0

THE VARIATIONAL PRINCIPLES OF MECHANICS, Cornelius Lanczos. Graduate level coverage of calculus of variations, equations of motion, relativistic mechanics, more. First inexpensive paperbound edition of classic treatise. Index. Bibliography. 418pp. 5⅜ x 8½. 65067-7

THE VARIOUS AND INGENIOUS MACHINES OF AGOSTINO RAMELLI: A Classic Sixteenth-Century Illustrated Treatise on Technology, Agostino Ramelli. One of the most widely known and copied works on machinery in the 16th century. 194 detailed plates of water pumps, grain mills, cranes, more. 608pp. 9 x 12. 28180-9

ORDINARY DIFFERENTIAL EQUATIONS AND STABILITY THEORY: An Introduction, David A. Sánchez. Brief, modern treatment. Linear equation, stability theory for autonomous and nonautonomous systems, etc. 164pp. 5⅜ x 8¼. 63828-6

ROTARY WING AERODYNAMICS, W. Z. Stepniewski. Clear, concise text cov-ers aerodynamic phenomena of the rotor and offers guidelines for helicopter perfor-mance evaluation. Originally prepared for NASA. 537 figures. 640pp. 6⅛ x 9¼. 64647-5

INTRODUCTION TO SPACE DYNAMICS, William Tyrrell Thomson. Com-prehensive, classic introduction to space-flight engineering for advanced undergrad-uate and graduate students. Includes vector algebra, kinematics, transformation of coordinates. Bibliography. Index. 352pp. 5⅜ x 8½. 65113-4

HISTORY OF STRENGTH OF MATERIALS, Stephen P. Timoshenko. Excellent historical survey of the strength of materials with many references to the theories of elasticity and structure. 245 figures. 452pp. 5⅜ x 8½. 61187-6

ANALYTICAL FRACTURE MECHANICS, David J. Unger. Self-contained text supplements standard fracture mechanics texts by focusing on analytical methods for determining crack-tip stress and strain fields. 336pp. 6⅛ x 9¼. 41737-9

Mathematics

HANDBOOK OF MATHEMATICAL FUNCTIONS WITH FORMULAS, GRAPHS, AND MATHEMATICAL TABLES, edited by Milton Abramowitz and Irene A. Stegun. Vast compendium: 29 sets of tables, some to as high as 20 places. 1,046pp. 8 x 10½. 61272-4

FUNCTIONAL ANALYSIS (Second Corrected Edition), George Bachman and Lawrence Narici. Excellent treatment of subject geared toward students with background in linear algebra, advanced calculus, physics and engineering. Text covers introduction to inner-product spaces, normed, metric spaces, and topological spaces; complete orthonormal sets, the Hahn-Banach Theorem and its consequences, and many other related subjects. 1966 ed. 544pp. 6⅛ x 9¼. 40251-7

ASYMPTOTIC EXPANSIONS OF INTEGRALS, Norman Bleistein & Richard A. Handelsman. Best introduction to important field with applications in a variety of scientific disciplines. New preface. Problems. Diagrams. Tables. Bibliography. Index. 448pp. 5⅜ x 8½. 65082-0

FAMOUS PROBLEMS OF GEOMETRY AND HOW TO SOLVE THEM, Benjamin Bold. Squaring the circle, trisecting the angle, duplicating the cube: learn their history, why they are impossible to solve, then solve them yourself. 128pp. 5⅜ x 8½. 24297-8

VECTOR AND TENSOR ANALYSIS WITH APPLICATIONS, A. I. Borisenko and I. E. Tarapov. Concise introduction. Worked-out problems, solutions, exercises. 257pp. 5⅝ x 8¼. 63833-2

THE ABSOLUTE DIFFERENTIAL CALCULUS (CALCULUS OF TENSORS), Tullio Levi-Civita. Great 20th-century mathematician's classic work on material necessary for mathematical grasp of theory of relativity. 452pp. 5⅝ x 8¼. 63401-9

AN INTRODUCTION TO ORDINARY DIFFERENTIAL EQUATIONS, Earl A. Coddington. A thorough and systematic first course in elementary differential equations for undergraduates in mathematics and science, with many exercises and problems (with answers). Index. 304pp. 5⅜ x 8½. 65942-9

FOURIER SERIES AND ORTHOGONAL FUNCTIONS, Harry F. Davis. An incisive text combining theory and practical example to introduce Fourier series, orthogonal functions and applications of the Fourier method to boundary-value problems. 570 exercises. Answers and notes. 416pp. 5⅜ x 8½. 65973-9

COMPUTABILITY AND UNSOLVABILITY, Martin Davis. Classic graduate-level introduction to theory of computability, usually referred to as theory of recurrent functions. New preface and appendix. 288pp. 5⅜ x 8½. 61471-9

ASYMPTOTIC METHODS IN ANALYSIS, N. G. de Bruijn. An inexpensive, comprehensive guide to asymptotic methods–the pioneering work that teaches by explaining worked examples in detail. Index. 224pp. 5⅜ x 8½ 64221-6

ESSAYS ON THE THEORY OF NUMBERS, Richard Dedekind. Two classic essays by great German mathematician: on the theory of irrational numbers; and on transfinite numbers and properties of natural numbers. 115pp. 5⅜ x 8½. 21010-3

APPLIED COMPLEX VARIABLES, John W. Dettman. Step-by-step coverage of fundamentals of analytic function theory—plus lucid exposition of five important applications: Potential Theory; Ordinary Differential Equations; Fourier Transforms; Laplace Transforms; Asymptotic Expansions. 66 figures. Exercises at chapter ends. 512pp. 5⅜ x 8½. 64670-X

INTRODUCTION TO LINEAR ALGEBRA AND DIFFERENTIAL EQUATIONS, John W. Dettman. Excellent text covers complex numbers, determinants, orthonormal bases, Laplace transforms, much more. Exercises with solutions. Undergraduate level. 416pp. 5⅜ x 8½. 65191-6

MATHEMATICAL METHODS IN PHYSICS AND ENGINEERING, John W. Dettman. Algebraically based approach to vectors, mapping, diffraction, other topics in applied math. Also generalized functions, analytic function theory, more. Exercises. 448pp. 5⅜ x 8¼. 65649-7

CALCULUS OF VARIATIONS WITH APPLICATIONS, George M. Ewing. Applications-oriented introduction to variational theory develops insight and promotes understanding of specialized books, research papers. Suitable for advanced undergraduate/graduate students as primary, supplementary text. 352pp. 5⅜ x 8½. 64856-7

COMPLEX VARIABLES, Francis J. Flanigan. Unusual approach, delaying complex algebra till harmonic functions have been analyzed from real variable viewpoint. Includes problems with answers. 364pp. 5⅜ x 8½. 61388-7

AN INTRODUCTION TO THE CALCULUS OF VARIATIONS, Charles Fox. Graduate-level text covers variations of an integral, isoperimetrical problems, least action, special relativity, approximations, more. References. 279pp. 5⅜ x 8½. 65499-0

CATASTROPHE THEORY FOR SCIENTISTS AND ENGINEERS, Robert Gilmore. Advanced-level treatment describes mathematics of theory grounded in the work of Poincaré, R. Thom, other mathematicians. Also important applications to problems in mathematics, physics, chemistry and engineering. 1981 edition. References. 28 tables. 397 black-and-white illustrations. xvii + 666pp. 6⅛ x 9¼. 67539-4

INTRODUCTION TO DIFFERENCE EQUATIONS, Samuel Goldberg. Exceptionally clear exposition of important discipline with applications to sociology, psychology, economics. Many illustrative examples; over 250 problems. 260pp. 5⅜ x 8½. 65084-7

NUMERICAL METHODS FOR SCIENTISTS AND ENGINEERS, Richard Hamming. Classic text stresses frequency approach in coverage of algorithms, polynomial approximation, Fourier approximation, exponential approximation, other topics. Revised and enlarged 2nd edition. 721pp. 5⅜ x 8½. 65241-6

INTRODUCTION TO NUMERICAL ANALYSIS (2nd Edition), F. B. Hildebrand. Classic, fundamental treatment covers computation, approximation, interpolation, numerical differentiation and integration, other topics. 150 new problems. 669pp. 5⅜ x 8½. 65363-3

THE FUNCTIONS OF MATHEMATICAL PHYSICS, Harry Hochstadt. Comprehensive treatment of orthogonal polynomials, hypergeometric functions, Hill's equation, much more. Bibliography. Index. 322pp. 5⅜ x 8½. 65214-9

THREE PEARLS OF NUMBER THEORY, A. Y. Khinchin. Three compelling puzzles require proof of a basic law governing the world of numbers. Challenges concern van der Waerden's theorem, the Landau-Schnirelmann hypothesis and Mann's theorem, and a solution to Waring's problem. Solutions included. 64pp. 5⅜ x 8½.
40026-3

CALCULUS REFRESHER FOR TECHNICAL PEOPLE, A. Albert Klaf. Covers important aspects of integral and differential calculus via 756 questions. 566 problems, most answered. 431pp. 5⅜ x 8½. 20370-0

THE PHILOSOPHY OF MATHEMATICS: An Introductory Essay, Stephan Körner. Surveys the views of Plato, Aristotle, Leibniz & Kant concerning propositions and theories of applied and pure mathematics. Introduction. Two appendices. Index. 198pp. 5⅜ x 8½. 25048-2

INTRODUCTORY REAL ANALYSIS, A.N. Kolmogorov, S. V. Fomin. Translated by Richard A. Silverman. Self-contained, evenly paced introduction to real and functional analysis. Some 350 problems. 403pp. 5⅜ x 8½. 61226-0

APPLIED ANALYSIS, Cornelius Lanczos. Classic work on analysis and design of finite processes for approximating solution of analytical problems. Algebraic equations, matrices, harmonic analysis, quadrature methods, much more. 559pp. 5⅜ x 8½.
65656-X

AN INTRODUCTION TO ALGEBRAIC STRUCTURES, Joseph Landin. Superb self-contained text covers "abstract algebra": sets and numbers, theory of groups, theory of rings, much more. Numerous well-chosen examples, exercises. 247pp. 5⅜ x 8½.
65940-2

SPECIAL FUNCTIONS, N. N. Lebedev. Translated by Richard Silverman. Famous Russian work treating more important special functions, with applications to specific problems of physics and engineering. 38 figures. 308pp. 5⅜ x 8½. 60624-4

QUALITATIVE THEORY OF DIFFERENTIAL EQUATIONS, V. V. Nemytskii and V.V. Stepanov. Classic graduate-level text by two prominent Soviet mathematicians covers classical differential equations as well as topological dynamics and ergodic theory. Bibliographies. 523pp. 5⅜ x 8½. 65954-2

NUMBER THEORY AND ITS HISTORY, Oystein Ore. Unusually clear, accessible introduction covers counting, properties of numbers, prime numbers, much more. Bibliography. 380pp. 5⅜ x 8½. 65620-9

THEORY OF MATRICES, Sam Perlis. Outstanding text covering rank, nonsingularity and inverses in connection with the development of canonical matrices under the relation of equivalence, and without the intervention of determinants. Includes exercises. 237pp. 5⅜ x 8½. 66810-X

CATALOG OF DOVER BOOKS

INTRODUCTION TO ANALYSIS, Maxwell Rosenlicht. Unusually clear, accessible coverage of set theory, real number system, metric spaces, continuous functions, Riemann integration, multiple integrals, more. Wide range of problems. Undergraduate level. Bibliography. 254pp. 5⅜ x 8½. 65038-3

MODERN NONLINEAR EQUATIONS, Thomas L. Saaty. Emphasizes practical solution of problems; covers seven types of equations. ". . . a welcome contribution to the existing literature...."–*Math Reviews.* 490pp. 5⅜ x 8½. 64232-1

MATRICES AND LINEAR ALGEBRA, Hans Schneider and George Phillip Barker. Basic textbook covers theory of matrices and its applications to systems of linear equations and related topics such as determinants, eigenvalues and differential equations. Numerous exercises. 432pp. 5⅜ x 8½. 66014-1

MATHEMATICS APPLIED TO CONTINUUM MECHANICS, Lee A. Segel. Analyzes models of fluid flow and solid deformation. For upper-level math, science and engineering students. 608pp. 5⅜ x 8½. 65369-2

ELEMENTS OF REAL ANALYSIS, David A. Sprecher. Classic text covers fundamental concepts, real number system, point sets, functions of a real variable, Fourier series, much more. Over 500 exercises. 352pp. 5⅜ x 8½. 65385-4

AN INTRODUCTION TO MATRICES, SETS AND GROUPS FOR SCIENCE STUDENTS, G. Stephenson. Concise, readable text introduces sets, groups, and most importantly, matrices to undergraduate students of physics, chemistry, and engineering. Problems. 164pp. 5⅜ x 8½. 65077-4

SET THEORY AND LOGIC, Robert R. Stoll. Lucid introduction to unified theory of mathematical concepts. Set theory and logic seen as tools for conceptual understanding of real number system. 496pp. 5⅜ x 8¼. 63829-4

TENSOR CALCULUS, J.L. Synge and A. Schild. Widely used introductory text covers spaces and tensors, basic operations in Riemannian space, non-Riemannian spaces, etc. 324pp. 5⅜ x 8¼. 63612-7

ORDINARY DIFFERENTIAL EQUATIONS, Morris Tenenbaum and Harry Pollard. Exhaustive survey of ordinary differential equations for undergraduates in mathematics, engineering, science. Thorough analysis of theorems. Diagrams. Bibliography. Index. 818pp. 5⅜ x 8½. 64940-7

INTEGRAL EQUATIONS, F. G. Tricomi. Authoritative, well-written treatment of extremely useful mathematical tool with wide applications. Volterra Equations, Fredholm Equations, much more. Advanced undergraduate to graduate level. Exercises. Bibliography. 238pp. 5⅜ x 8½. 64828-1

FOURIER SERIES, Georgi P. Tolstov. Translated by Richard A. Silverman. A valuable addition to the literature on the subject, moving clearly from subject to subject and theorem to theorem. 107 problems, answers. 336pp. 5⅜ x 8½. 63317-9

POPULAR LECTURES ON MATHEMATICAL LOGIC, Hao Wang. Noted logician's lucid treatment of historical developments, set theory, model theory, recursion theory and constructivism, proof theory, more. 3 appendixes. Bibliography. 1981 edition. ix + 283pp. 5⅜ x 8½. 67632-3

CALCULUS OF VARIATIONS, Robert Weinstock. Basic introduction covering isoperimetric problems, theory of elasticity, quantum mechanics, electrostatics, etc. Exercises throughout. 326pp. 5⅜ x 8½. 63069-2

THE CONTINUUM: A Critical Examination of the Foundation of Analysis, Hermann Weyl. Classic of 20th-century foundational research deals with the conceptual problem posed by the continuum. 156pp. 5⅜ x 8½. 67982-9

CHALLENGING MATHEMATICAL PROBLEMS WITH ELEMENTARY SOLUTIONS, A. M. Yaglom and I. M. Yaglom. Over 170 challenging problems on probability theory, combinatorial analysis, points and lines, topology, convex polygons, many other topics. Solutions. Total of 445pp. 5⅜ x 8½. Two-vol. set.
Vol. I: 65536-9 Vol. II: 65537-7

A SURVEY OF NUMERICAL MATHEMATICS, David M. Young and Robert Todd Gregory. Broad self-contained coverage of computer-oriented numerical algorithms for solving various types of mathematical problems in linear algebra, ordinary and partial, differential equations, much more. Exercises. Total of 1,248pp. 5⅜ x 8½. Two volumes. Vol. I: 65691-8 Vol. II: 65692-6

INTRODUCTION TO PARTIAL DIFFERENTIAL EQUATIONS WITH APPLICATIONS, E. C. Zachmanoglou and Dale W. Thoe. Essentials of partial differential equations applied to common problems in engineering and the physical sciences. Problems and answers. 416pp. 5⅜ x 8½. 65251-3

THE THEORY OF GROUPS, Hans J. Zassenhaus. Well-written graduate-level text acquaints reader with group-theoretic methods and demonstrates their usefulness in mathematics. Axioms, the calculus of complexes, homomorphic mapping, p-group theory, more. Many proofs shorter and more transparent than older ones. 276pp. 5⅜ x 8½. 40922-8

DISTRIBUTION THEORY AND TRANSFORM ANALYSIS: An Introduction to Generalized Functions, with Applications, A. H. Zemanian. Provides basics of distribution theory, describes generalized Fourier and Laplace transformations. Numerous problems. 384pp. 5⅜ x 8½. 65479-6

Math–Decision Theory, Statistics, Probability

ELEMENTARY DECISION THEORY, Herman Chernoff and Lincoln E. Moses. Clear introduction to statistics and statistical theory covers data processing, probability and random variables, testing hypotheses, much more. Exercises. 364pp. 5⅜ x 8½. 65218-1

CATALOG OF DOVER BOOKS

STATISTICS MANUAL, Edwin L. Crow et al. Comprehensive, practical collection of classical and modern methods prepared by U.S. Naval Ordnance Test Station. Stress on use. Basics of statistics assumed. 288pp. 5⅜ x 8½.   60599-X

SOME THEORY OF SAMPLING, William Edwards Deming. Analysis of the problems, theory and design of sampling techniques for social scientists, industrial managers and others who find statistics important at work. 61 tables. 90 figures. xvii +602pp. 5⅜ x 8½.   64684-X

STATISTICAL ADJUSTMENT OF DATA, W. Edwards Deming. Introduction to basic concepts of statistics, curve fitting, least squares solution, conditions without parameter, conditions containing parameters. 26 exercises worked out. 271pp. 5⅜ x 8½.   64685-8

LINEAR PROGRAMMING AND ECONOMIC ANALYSIS, Robert Dorfman, Paul A. Samuelson and Robert M. Solow. First comprehensive treatment of linear programming in standard economic analysis. Game theory, modern welfare economics, Leontief input-output, more. 525pp. 5⅜ x 8½.   65491-5

DICTIONARY/OUTLINE OF BASIC STATISTICS, John E. Freund and Frank J. Williams. A clear concise dictionary of over 1,000 statistical terms and an outline of statistical formulas covering probability, nonparametric tests, much more. 208pp. 5⅜ x 8½.   66796-0

PROBABILITY: An Introduction, Samuel Goldberg. Excellent basic text covers set theory, probability theory for finite sample spaces, binomial theorem, much more. 360 problems. Bibliographies. 322pp. 5⅜ x 8½.   65252-1

GAMES AND DECISIONS: Introduction and Critical Survey, R. Duncan Luce and Howard Raiffa. Superb nontechnical introduction to game theory, primarily applied to social sciences. Utility theory, zero-sum games, n-person games, decision-making, much more. Bibliography. 509pp. 5⅜ x 8½.   65943-7

FIFTY CHALLENGING PROBLEMS IN PROBABILITY WITH SOLUTIONS, Frederick Mosteller. Remarkable puzzlers, graded in difficulty, illustrate elementary and advanced aspects of probability. Detailed solutions. 88pp. 5⅜ x 8½.   65355-2

PROBABILITY THEORY: A Concise Course, Y. A. Rozanov. Highly readable, self-contained introduction covers combination of events, dependent events, Bernoulli trials, etc. 148pp. 5⅜ x 8¼.   63544-9

STATISTICAL METHOD FROM THE VIEWPOINT OF QUALITY CONTROL, Walter A. Shewhart. Important text explains regulation of variables, uses of statistical control to achieve quality control in industry, agriculture, other areas. 192pp. 5⅜ x 8½.   65232-7

THE COMPLEAT STRATEGYST: Being a Primer on the Theory of Games of Strategy, J. D. Williams. Highly entertaining classic describes, with many illustrated examples, how to select best strategies in conflict situations. Prefaces. Appendices. 268pp. 5⅜ x 8½.   25101-2

Math–Geometry and Topology

ELEMENTARY CONCEPTS OF TOPOLOGY, Paul Alexandroff. Elegant, intuitive approach to topology from set-theoretic topology to Betti groups; how concepts of topology are useful in math and physics. 25 figures. 57pp. 5⅜ x 8½. 60747-X

COMBINATORIAL TOPOLOGY, P. S. Alexandrov. Clearly written, well-organized, three-part text begins by dealing with certain classic problems without using the formal techniques of homology theory and advances to the central concept, the Betti groups. Numerous detailed examples. 654pp. 5⅜ x 8½. 40179-0

EXPERIMENTS IN TOPOLOGY, Stephen Barr. Classic, lively explanation of one of the byways of mathematics. Klein bottles, Moebius strips, projective planes, map coloring, problem of the Koenigsberg bridges, much more, described with clarity and wit. 43 figures. 210pp. 5⅜ x 8½. 25933-1

CONFORMAL MAPPING ON RIEMANN SURFACES, Harvey Cohn. Lucid, insightful book presents ideal coverage of subject. 334 exercises make book perfect for self-study. 55 figures. 352pp. 5⅝ x 8¼. 64025-6

THE GEOMETRY OF RENÉ DESCARTES, René Descartes. The great work founded analytical geometry. Original French text, Descartes's own diagrams, together with definitive Smith-Latham translation. 244pp. 5⅜ x 8½. 60068-8

THE THIRTEEN BOOKS OF EUCLID'S ELEMENTS, translated with introduction and commentary by Sir Thomas L. Heath. Definitive edition. Textual and linguistic notes, mathematical analysis. 2,500 years of critical commentary. Unabridged. 1,414pp. 5⅜ x 8½.. Three-vol. set.
Vol. I: 60088-2 Vol. II: 60089-0 Vol. III: 60090-4

GEOMETRY OF COMPLEX NUMBERS, Hans Schwerdtfeger. Illuminating, widely praised book on analytic geometry of circles, the Moebius transformation, and two-dimensional non-Euclidean geometries. 200pp. 5⅝ x 8¼. 63830-8

DIFFERENTIAL GEOMETRY, Heinrich W. Guggenheimer. Local differential geometry as an application of advanced calculus and linear algebra. Curvature, transformation groups, surfaces, more. Exercises. 62 figures. 378pp. 5⅜ x 8½. 63433-7

CURVATURE AND HOMOLOGY: Enlarged Edition, Samuel I. Goldberg. Revised edition examines topology of differentiable manifolds; curvature, homology of Riemannian manifolds; compact Lie groups; complex manifolds; curvature, homology of Kaehler manifolds. New Preface. Four new appendixes. 416pp. 5⅜ x 8½. 40207-X

TOPOLOGY, John G. Hocking and Gail S. Young. Superb one-year course in classical topology. Topological spaces and functions, point-set topology, much more. Examples and problems. Bibliography. Index. 384pp. 5⅝ x 8¼. 65676-4

LECTURES ON CLASSICAL DIFFERENTIAL GEOMETRY, Second Edition, Dirk J. Struik. Excellent brief introduction covers curves, theory of surfaces, fundamental equations, geometry on a surface, conformal mapping, other topics. Problems. 240pp. 5⅜ x 8½. 65609-8

Math–History of

A SHORT ACCOUNT OF THE HISTORY OF MATHEMATICS, W. W. Rouse Ball. One of clearest, most authoritative surveys from the Egyptians and Phoenicians through 19th-century figures such as Grassman, Galois, Riemann. Fourth edition. 522pp. 5⅜ x 8½. 20630-0

THE HISTORY OF THE CALCULUS AND ITS CONCEPTUAL DEVELOPMENT, Carl B. Boyer. Origins in antiquity, medieval contributions, work of Newton, Leibniz, rigorous formulation. Treatment is verbal. 346pp. 5⅜ x 8½. 60509-4

THE HISTORICAL ROOTS OF ELEMENTARY MATHEMATICS, Lucas N. H. Bunt, Phillip S. Jones, and Jack D. Bedient. Fundamental underpinnings of modern arithmetic, algebra, geometry and number systems derived from ancient civilizations. 320pp. 5⅜ x 8½. 25563-8

A HISTORY OF MATHEMATICAL NOTATIONS, Florian Cajori. This classic study notes the first appearance of a mathematical symbol and its origin, the competition it encountered, its spread among writers in different countries, its rise to popularity, its eventual decline or ultimate survival. Original 1929 two-volume edition presented here in one volume. xxviii+820pp. 5⅜ x 8½. 67766-4

GAMES, GODS & GAMBLING: A History of Probability and Statistical Ideas, F. N. David. Episodes from the lives of Galileo, Fermat, Pascal, and others illustrate this fascinating account of the roots of mathematics. Features thought-provoking references to classics, archaeology, biography, poetry. 1962 edition. 304pp. 5⅜ x 8½. (Available in U.S. only.) 40023-9

OF MEN AND NUMBERS: The Story of the Great Mathematicians, Jane Muir. Fascinating accounts of the lives and accomplishments of history's greatest mathematical minds–Pythagoras, Descartes, Euler, Pascal, Cantor, many more. Anecdotal, illuminating. 30 diagrams. Bibliography. 256pp. 5⅜ x 8½. 28973-7

HISTORY OF MATHEMATICS, David E. Smith. Nontechnical survey from ancient Greece and Orient to late 19th century; evolution of arithmetic, geometry, trigonometry, calculating devices, algebra, the calculus. 362 illustrations. 1,355pp. 5⅜ x 8½. Two-vol. set. Vol. I: 20429-4 Vol. II: 20430-8

A CONCISE HISTORY OF MATHEMATICS, Dirk J. Struik. The best brief history of mathematics. Stresses origins and covers every major figure from ancient Near East to 19th century. 41 illustrations. 195pp. 5⅜ x 8½. 60255-9

Physics

OPTICAL RESONANCE AND TWO-LEVEL ATOMS, L. Allen and J. H. Eberly. Clear, comprehensive introduction to basic principles behind all quantum optical resonance phenomena. 53 illustrations. Preface. Index. 256pp. 5⅜ x 8½.   65533-4

ULTRASONIC ABSORPTION: An Introduction to the Theory of Sound Absorption and Dispersion in Gases, Liquids and Solids, A. B. Bhatia. Standard reference in the field provides a clear, systematically organized introductory review of fundamental concepts for advanced graduate students, research workers. Numerous diagrams. Bibliography. 440pp. 5⅜ x 8½.   64917-2

QUANTUM THEORY, David Bohm. This advanced undergraduate-level text presents the quantum theory in terms of qualitative and imaginative concepts, followed by specific applications worked out in mathematical detail. Preface. Index. 655pp. 5⅜ x 8½.   65969-0

ATOMIC PHYSICS (8th edition), Max Born. Nobel laureate's lucid treatment of kinetic theory of gases, elementary particles, nuclear atom, wave-corpuscles, atomic structure and spectral lines, much more. Over 40 appendices, bibliography. 495pp. 5⅜ x 8½.   65984-4

AN INTRODUCTION TO HAMILTONIAN OPTICS, H. A. Buchdahl. Detailed account of the Hamiltonian treatment of aberration theory in geometrical optics. Many classes of optical systems defined in terms of the symmetries they possess. Problems with detailed solutions. 1970 edition. xv + 360pp. 5⅜ x 8½.   67597-1

THIRTY YEARS THAT SHOOK PHYSICS: The Story of Quantum Theory, George Gamow. Lucid, accessible introduction to influential theory of energy and matter. Careful explanations of Dirac's anti-particles, Bohr's model of the atom, much more. 12 plates. Numerous drawings. 240pp. 5⅜ x 8½.   24895-X

ELECTRONIC STRUCTURE AND THE PROPERTIES OF SOLIDS: The Physics of the Chemical Bond, Walter A. Harrison. Innovative text offers basic understanding of the electronic structure of covalent and ionic solids, simple metals, transition metals and their compounds. Problems. 1980 edition. 582pp. 6⅛ x 9¼.
66021-4

HYDRODYNAMIC AND HYDROMAGNETIC STABILITY, S. Chandrasekhar. Lucid examination of the Rayleigh-Benard problem; clear coverage of the theory of instabilities causing convection. 704pp. 5⅜ x 8¼.   64071-X

INVESTIGATIONS ON THE THEORY OF THE BROWNIAN MOVEMENT, Albert Einstein. Five papers (1905–8) investigating dynamics of Brownian motion and evolving elementary theory. Notes by R. Fürth. 122pp. 5⅜ x 8½.   60304-0

THE PHYSICS OF WAVES, William C. Elmore and Mark A. Heald. Unique overview of classical wave theory. Acoustics, optics, electromagnetic radiation, more. Ideal as classroom text or for self-study. Problems. 477pp. 5⅜ x 8½.   64926-1

CATALOG OF DOVER BOOKS

PHYSICAL PRINCIPLES OF THE QUANTUM THEORY, Werner Heisenberg. Nobel Laureate discusses quantum theory, uncertainty, wave mechanics, work of Dirac, Schroedinger, Compton, Wilson, Einstein, etc. 184pp. 5⅜ x 8½.   60113-7

ATOMIC SPECTRA AND ATOMIC STRUCTURE, Gerhard Herzberg. One of best introductions; especially for specialist in other fields. Treatment is physical rather than mathematical. 80 illustrations. 257pp. 5⅜ x 8½.   60115-3

AN INTRODUCTION TO STATISTICAL THERMODYNAMICS, Terrell L. Hill. Excellent basic text offers wide-ranging coverage of quantum statistical mechanics, systems of interacting molecules, quantum statistics, more. 523pp. 5⅜ x 8½.
65242-4

THEORETICAL PHYSICS, Georg Joos, with Ira M. Freeman. Classic overview covers essential math, mechanics, electromagnetic theory, thermodynamics, quantum mechanics, nuclear physics, other topics. First paperback edition. xxiii + 885pp. 5⅜ x 8½.   65227-0

PROBLEMS AND SOLUTIONS IN QUANTUM CHEMISTRY AND PHYSICS, Charles S. Johnson, Jr. and Lee G. Pedersen. Unusually varied problems, detailed solutions in coverage of quantum mechanics, wave mechanics, angular momentum, molecular spectroscopy, more. 280 problems plus 139 supplementary exercises. 430pp. 6½ x 9¼.   65236-X

THEORETICAL SOLID STATE PHYSICS, Vol. 1: Perfect Lattices in Equilibrium; Vol. II: Non-Equilibrium and Disorder, William Jones and Norman H. March. Monumental reference work covers fundamental theory of equilibrium properties of perfect crystalline solids, non-equilibrium properties, defects and disordered systems. Appendices. Problems. Preface. Diagrams. Index. Bibliography. Total of 1,301pp. 5⅜ x 8½. Two volumes.   Vol. I: 65015-4 Vol. II: 65016-2

A TREATISE ON ELECTRICITY AND MAGNETISM, James Clerk Maxwell. Important foundation work of modern physics. Brings to final form Maxwell's theory of electromagnetism and rigorously derives his general equations of field theory. 1,084pp. 5⅜ x 8½. Two-vol. set.   Vol. I: 60636-8 Vol. II: 60637-6

OPTICKS, Sir Isaac Newton. Newton's own experiments with spectroscopy, colors, lenses, reflection, refraction, etc., in language the layman can follow. Foreword by Albert Einstein. 532pp. 5⅜ x 8½.   60205-2

THEORY OF ELECTROMAGNETIC WAVE PROPAGATION, Charles Herach Papas. Graduate-level study discusses the Maxwell field equations, radiation from wire antennas, the Doppler effect and more. xiii + 244pp. 5⅜ x 8½.   65678-5

INTRODUCTION TO QUANTUM MECHANICS With Applications to Chemistry, Linus Pauling & E. Bright Wilson, Jr. Classic undergraduate text by Nobel Prize winner applies quantum mechanics to chemical and physical problems. Numerous tables and figures enhance the text. Chapter bibliographies. Appendices. Index. 468pp. 5⅜ x 8½.   64871-0

CATALOG OF DOVER BOOKS

METHODS OF THERMODYNAMICS, Howard Reiss. Outstanding text focuses on physical technique of thermodynamics, typical problem areas of understanding, and significance and use of thermodynamic potential. 1965 edition. 238pp. 5⅜ x 8½. 69445-3

TENSOR ANALYSIS FOR PHYSICISTS, J. A. Schouten. Concise exposition of the mathematical basis of tensor analysis, integrated with well-chosen physical examples of the theory. Exercises. Index. Bibliography. 289pp. 5⅜ x 8½. 65582-2

RELATIVITY IN ILLUSTRATIONS, Jacob T. Schwartz. Clear nontechnical treatment makes relativity more accessible than ever before. Over 60 drawings illustrate concepts more clearly than text alone. Only high school geometry needed. Bibliography. 128pp. 6⅛ x 9¼. 25965-X

THE ELECTROMAGNETIC FIELD, Albert Shadowitz. Comprehensive undergraduate text covers basics of electric and magnetic fields, builds up to electromagnetic theory. Also related topics, including relativity. Over 900 problems. 768pp. 5⅝ x 8¼. 65660-8

GREAT EXPERIMENTS IN PHYSICS: Firsthand Accounts from Galileo to Einstein, edited by Morris H. Shamos. 25 crucial discoveries: Newton's laws of motion, Chadwick's study of the neutron, Hertz on electromagnetic waves, more. Original accounts clearly annotated. 370pp. 5⅜ x 8½. 25346-5

RELATIVITY, THERMODYNAMICS AND COSMOLOGY, Richard C. Tolman. Landmark study extends thermodynamics to special, general relativity; also applications of relativistic mechanics, thermodynamics to cosmological models. 501pp. 5⅜ x 8½. 65383-8

LIGHT SCATTERING BY SMALL PARTICLES, H. C. van de Hulst. Comprehensive treatment including full range of useful approximation methods for researchers in chemistry, meteorology and astronomy. 44 illustrations. 470pp. 5⅜ x 8½. 64228-3

STATISTICAL PHYSICS, Gregory H. Wannier. Classic text combines thermodynamics, statistical mechanics and kinetic theory in one unified presentation of thermal physics. Problems with solutions. Bibliography. 532pp. 5⅜ x 8½. 65401-X